PRAISE FOR KIANNAH FROST

From *"Breaking the Chains of Fatphobia:"*

This book is a beacon of hope in the fight against fatphobia. It's an essential read for anyone looking to understand and challenge weight-based discrimination. Get ready to be inspired and join the movement!

— ELVIRA LINDSEY

From *"Fearless Connection:"*

This 3-in-1 collection has been a game-changer in my personal growth journey. Each book provided valuable insights and tools to help me transform my life.

— JANELLE PRINCE

From *"The Power of Psychological Safety:"*

An essential read for today's leaders. This book provides the tools and knowledge to create a supportive, high-performing work culture.

— JESSE WILKINSON

INFOTAINMENT UNLEASHED

INFOTAINMENT UNLEASHED

EXPLORING THE EVOLUTION, IMPACT, AND FUTURE OF INTEGRATED ENTERTAINMENT AND INFORMATION SYSTEMS

NAVIGATING THE DIGITAL AGE

KIANNAH FROST

Book
Bound Studios

To the curious minds and the relentless innovators, this book is dedicated to you. As we navigate the ever-evolving landscape of infotainment, it is your passion and creativity that continue to push the boundaries of what is possible. May this book serve as a guide and inspiration for those seeking to understand and contribute to the exciting world of infotainment. Together, let us embrace the infotainment revolution and shape the future of how we connect, entertain, and inform ourselves.

In the age of information, ignorance is a choice.

— DONNY MILLER

CONTENTS

INTRODUCTION TO INFOTAINMENT

In the age of information, the world has witnessed a remarkable transformation in the way we consume and process content. The rapid evolution of technology and the ever-growing influence of social media have given birth to a new era - the era of infotainment. This exciting and dynamic landscape has blurred the lines between information and entertainment, creating a unique blend that captivates audiences and keeps them hooked.

Infotainment has become an integral part of our daily lives as we constantly seek to be informed and entertained simultaneously. From news channels to social media platforms, infotainment has permeated every aspect of our lives, shaping the way we perceive and interact with the world around us. It has redefined the boundaries of traditional media, paving the way for innovative formats that cater to the diverse needs and preferences of modern audiences.

As we stand at the dawn of the infotainment era, it is crucial to understand the forces that drive this phenomenon and the impact it has on our society. This book aims to provide a comprehensive overview of the world of infotainment, delving into its origins, its various forms, and its challenges and opportunities. By exploring the multifaceted landscape of infotainment, we hope to equip readers with the knowledge and insights necessary to navigate this ever-evolving domain.

So, let us embark on this fascinating journey together as we unravel the mysteries of infotainment and discover the myriad ways in which it is transforming our lives. Welcome to the dawn of the infotainment era – a world where information and entertainment collide, creating a captivating and immersive experience that is reshaping the very fabric of our society.

Tracing the Roots of Infotainment

The term "infotainment" is a portmanteau of the words "information" and "entertainment," and it aptly describes the phenomenon of combining these two elements to create content that is both informative and engaging. To truly understand the roots of infotainment, we must first delve into the history of information dissemination and entertainment in various forms of media.

The concept of infotainment can be traced back to the early days of print media when newspapers began incorporating human interest stories and comic strips alongside hard news to appeal to a wider audience. As radio emerged as a popular medium in the early 20th century, news broadcasts were often interspersed with music and other forms of entertainment to keep listeners engaged. The advent of television further expanded the possibilities for infotainment, as news programs began to include features, interviews, and other entertaining segments to maintain viewer interest.

However, the true explosion of infotainment as we know it today can be attributed to the rise of cable television and the internet. The 24-hour news cycle and the proliferation of niche channels catering to specific interests increased the demand for content that could inform and entertain. This gave birth to an array of infotainment programs, ranging from talk shows and documentaries to reality TV and investigative journalism.

The internet has further revolutionized the infotainment landscape, as it has provided a platform for content creators to reach a global audience with ease. Social media, blogs, podcasts, and video-sharing platforms like YouTube have given rise to a new generation of infotainment stars who blend humor, storytelling, and information to engage their audiences. These digital platforms have also allowed for greater interactivity and personalization, enabling users to tailor their infotainment experiences to their specific interests and preferences.

As we trace the roots of infotainment, it becomes evident that the desire for engaging, informative content has been a constant throughout the evolution of media. What has changed, however, is the way in which this content is delivered and consumed. The rise of digital platforms has democratized the infotainment landscape, allowing for greater diversity and creativity in the content that is produced and shared. As we continue to explore the world of infotainment, it is important to recognize and appreciate the rich history that has shaped this fascinating phenomenon.

Navigating the World of Infotainment

The primary purpose of this book is to serve as a comprehensive guide for readers who wish to delve into the fascinating world of infotainment. As an industry expert, I have observed the rapid evolution of this field and its growing impact on various aspects of our lives. This book aims to provide a thorough understanding of the concept of infotainment, its origins, and its current manifestations in different industries.

This book is designed to cater to a diverse audience, including media professionals, students, educators, and anyone with a keen interest in understanding the convergence of information and entertainment. The book will equip readers with the knowledge and tools necessary to critically analyze and appreciate the various forms of infotainment that surround us.

In this age of information overload, it is crucial to understand the role that infotainment plays in shaping our perceptions, opinions, and decisions. This book will help readers develop a discerning eye, enabling them to differentiate between valuable information and mere distractions. By understanding the mechanics of infotainment, readers will be better equipped to navigate the complex media landscape and make informed choices about the content they consume.

Furthermore, this book will provide insights into the creative and business aspects of the infotainment industry, offering valuable advice for those who aspire to build a career in this field. Drawing from my own experiences and expertise, I will share practical tips and strategies for success in the ever-evolving world of infotainment.

In essence, this book is a comprehensive resource that aims to educate, inform, and inspire readers to embrace the infotainment revolution and make the most of the opportunities it presents.

Insights from an Industry Expert

As an author and industry expert, I have had the privilege of witnessing the evolution of infotainment firsthand. From the early days of radio and television to the current digital age, the landscape of infotainment has transformed dramatically. This transformation has not only impacted the way we consume information and entertainment but has also shaped our society in ways we could never have imagined.

Throughout my career, I have had the opportunity to work with various media organizations, technology companies, and content creators. This experience has provided me with a unique vantage point to observe the trends and developments in the infotainment industry. I have seen the rise of new platforms, the convergence of media, and the emergence of innovative storytelling techniques. These insights have allowed me to develop a deep understanding of the challenges and opportunities that lie ahead in the world of infotainment.

One of the key insights I have gained over the years is the importance of striking the right balance between information and entertainment. Infotainment, at its core, is a fusion of these two elements. However, it is essential to ensure that the information presented is accurate, relevant, and reliable while still engaging and entertaining. This delicate balance is what sets successful infotainment apart from mere sensationalism or dry, uninteresting content.

Another crucial aspect of infotainment is the role of technology in shaping the industry. As an industry expert, I have seen how technological advancements have revolutionized the way we create, distribute, and consume infotainment. From the rise of social media and streaming platforms to the development of virtual and augmented reality, technology has opened up new avenues for infotainment to thrive. Content creators and media organizations need to stay ahead of the curve and adapt to these changes to remain relevant in this ever-evolving landscape.

Lastly, I believe that the future of infotainment lies in its ability to foster a sense of global connectedness. In today's world, where information is readily available at our fingertips, infotainment has the power to bridge cultural, geographical, and linguistic divides. By presenting information in an engaging and entertaining manner, infotainment can help us understand and appreciate the diverse perspectives and experiences of people from around the world.

As we delve deeper into the world of infotainment in this book, I hope to share my insights and experiences as an industry expert to help you navigate this fascinating landscape. Together, we will explore the multifaceted world of infotainment and embrace the revolution that is transforming the way we consume information and entertainment.

Embracing the Infotainment Revolution

As we reach the end of this introductory chapter, it is essential to recognize the profound impact that infotainment has had on our society. The fusion of information and entertainment has revolutionized the way we consume content, engage with the world around us, and even perceive reality. Embracing the infotainment revolution is not only about acknowledging its existence but also understanding its implications and harnessing its potential for the betterment of our lives.

The infotainment era has undoubtedly brought about a paradigm shift in the media landscape. Traditional forms of media have been forced to adapt and evolve to stay relevant in this ever-changing environment. As a result, we have witnessed the rise of new media platforms, innovative storytelling techniques, and a plethora of content that caters to diverse interests and preferences. This transformation has expanded the horizons of creativity and democratized access to information and entertainment for millions of people across the globe.

However, the infotainment revolution is not without its challenges. The blurring of lines between fact and fiction, the proliferation of sensationalism, and the potential for manipulation and misinformation are some of the pressing concerns that need to be addressed. As industry experts, content creators, and consumers, we must strive to strike a balance between the allure of entertainment and the responsibility of providing accurate and reliable information. This involves fostering media literacy, promoting ethical journalism, and cultivating a discerning audience that can navigate the complex world of infotainment with ease and confidence.

In the subsequent chapters of this book, we will delve deeper into the various facets of the infotainment landscape, exploring its history, evolution, and impact on different sectors such as news, education, and advertising. We will also examine the role of technology in shaping the future of infotainment and discuss the emerging trends that are set to redefine the way we consume and engage with content.

As an author and industry expert, my goal is to provide you with valuable insights, thought-provoking discussions, and practical guidance that will help you navigate the world of infotainment with ease and confidence. By understanding the nuances of this fascinating domain, we can collectively embrace the infotainment revolution and harness its potential to inform, educate, and entertain in a responsible and meaningful manner.

So, let us embark on this exciting journey together and explore the captivating realm of infotainment, where information and entertainment converge to create a dynamic, engaging, and enriching experience for all.

1

EVOLUTION OF INFOTAINMENT SYSTEMS

A vintage car cruising down a neon-lit street in the 1980s.

Infotainment, a term coined by combining 'information' and 'entertainment,' has become an integral part of our daily lives. From the moment we wake up to the time we go to bed, we are constantly surrounded by various forms of infotainment. Be it through our smartphones, televisions, or in-car systems, infotainment has revolutionized the way we consume information and entertainment. This chapter delves into the fascinating journey of infotainment systems, tracing their evolution from the early days of innovation to the cutting-edge technology we experience today.

The concept of infotainment is not new; it has existed for centuries in various forms. However, the rapid technological advancements over the past few decades have significantly transformed the landscape of infotainment systems. The emergence of digital technology, in particular, has played a pivotal role in shaping the modern infotainment systems we have come to know and rely on.

As we explore the evolution of infotainment systems, we will closely examine the early innovations and pioneers who laid the groundwork for this industry. We will examine the transformative impact of digital technology on infotainment systems and how it has enabled the seamless integration of these systems into our modern lives. Furthermore, we will discuss the future of infotainment, highlighting the trends and predictions that are expected to shape the industry in the coming years.

The ongoing evolution of infotainment systems is a testament to human ingenuity and our insatiable appetite for information and entertainment. As we continue to push the boundaries of technology, infotainment systems will undoubtedly play an even more significant role in our lives, making the world a more connected, informed, and entertaining place to live in.

The Emergence of Infotainment: Early Innovations and Pioneers

As we delve into the fascinating world of infotainment, it is essential to understand its roots and early beginnings. The term "infotainment" is a portmanteau of "information" and "entertainment," which aptly describes the purpose of these systems: to inform and entertain simultaneously. In this section, we will explore the emergence of infotainment, highlighting the early innovations and pioneers that paved the way for the sophisticated systems we enjoy today.

The concept of infotainment can be traced back to the early days of mass media when newspapers, radio, and television began to blend information with entertainment. One of the first examples of infotainment was the newspaper comic strip, which combined humor and storytelling with current events and social commentary. This innovative format allowed readers to engage with the news in a more enjoyable and accessible way.

Radio soon followed suit, with programs such as Orson Welles' "The War of the Worlds" broadcast in 1938, which blurred the lines between fiction and reality, creating a captivating and immersive experience for listeners. As television emerged as a dominant medium, shows like "The Twilight Zone" and "60 Minutes" continued to push the boundaries of infotainment, offering thought-provoking content that both informed and entertained viewers.

The 1980s and 1990s saw the rise of cable television and the proliferation of specialized channels, which further fueled the growth of infotainment. Channels like MTV, ESPN, and CNN combined entertainment and information in innovative ways, catering to niche audiences and interests. This era also witnessed the birth of the 24-hour news cycle, which placed a premium on engaging and attention-grabbing content to keep viewers tuned in.

One of the most significant early pioneers of infotainment was Ted Turner, who founded CNN in 1980. Turner's vision was to create a news network that would provide continuous coverage of global events, with a mix of hard news, analysis, and human-interest stories. This groundbreaking approach revolutionized the way people consumed news and set the stage for the modern infotainment landscape.

Another notable figure in the evolution of infotainment is Oprah Winfrey, whose eponymous talk show debuted in 1986. "The Oprah Winfrey Show" skillfully blended celebrity interviews, personal stories, and expert advice, creating a unique and compelling format that resonated with millions of viewers. Winfrey's ability to connect with her audience and tackle complex issues with empathy and insight made her a trailblazer in the world of infotainment.

In summary, the emergence of infotainment can be attributed to the innovative efforts of early pioneers who recognized the potential of combining information and entertainment to create engaging and accessible content. From newspaper comic strips to groundbreaking television programs and networks, these early innovations laid the foundation for the sophisticated infotainment systems that have become an integral part of our modern lives.

The Rise of Digital Technology: Transforming Infotainment Systems

The advent of digital technology has played a pivotal role in the transformation of infotainment systems, revolutionizing the way we consume information and entertainment. This section will delve into the key developments and breakthroughs that have shaped the modern infotainment landscape, highlighting the impact of digital technology on various aspects of our lives.

The shift from analog to digital technology has been a game-changer for infotainment systems. In the early days, infotainment relied heavily on traditional media formats such as print, radio, and television. However, the introduction of digital technology has opened up new avenues for content creation, distribution, and consumption. The emergence of the internet, in particular, has had a profound impact on the way we access and engage with infotainment.

One of the most significant developments in the digital era has been the proliferation of online platforms and streaming services. These platforms have enabled users to access a vast array of content at their fingertips, ranging from news articles and podcasts to movies and music. The convenience and accessibility offered by these services have led to a surge in demand for infotainment, with consumers now able to access their favorite content anytime, anywhere.

Another key development in the digital transformation of infotainment systems has been the rise of social media. Platforms such as Facebook, Twitter, and Instagram have not only changed the way we communicate with one another but have also become important sources of infotainment. Social media has made it possible for users to share and discover new content while also providing a platform for content creators to reach a wider audience. The real-time nature of social media has also made it an invaluable tool for staying up-to-date with the latest news and events.

The evolution of digital technology has also led to the emergence of new devices and gadgets that have further enhanced the infotainment experience. Smartphones and tablets, for instance, have become indispensable tools for accessing infotainment on the go. These devices have also paved the way for developing various apps and software that cater to different infotainment needs, from news aggregators and e-readers to video streaming and gaming apps.

In addition to portable devices, digital technology has also transformed infotainment systems in our homes and vehicles. Smart TVs and voice-activated assistants like Amazon's Alexa and Google Home have made it easier than ever to access and control infotainment content, while in-car infotainment systems have evolved to include features such as GPS navigation, hands-free calling, and smartphone integration.

In conclusion, the rise of digital technology has transformed infotainment systems, reshaping how we access and engage with content. The integration of digital technology into various aspects of our lives has made infotainment more accessible, convenient, and personalized than ever before. As digital technology continues to advance, it is likely that infotainment systems will continue to evolve, offering even more immersive and engaging experiences for users.

The Integration of Infotainment Systems into Modern Life

As infotainment systems have evolved, they have become an integral part of our daily lives, seamlessly blending into various aspects of modern society. This integration has been facilitated by advancements in digital technology, which have allowed infotainment systems to become more accessible, user-friendly, and versatile. This section will explore how infotainment systems have permeated various aspects of our lives, from our homes and workplaces to our vehicles and public spaces.

One of the most significant ways in which infotainment systems have integrated into modern life is through the proliferation of smartphones and mobile devices. These devices have become our constant companions, providing instant access to a wealth of information and entertainment options at our fingertips. From streaming music and videos to browsing social media and playing games, smartphones have revolutionized the way we consume and interact with infotainment content.

In addition to smartphones, infotainment systems have also become a staple in our homes through the widespread adoption of smart TVs and streaming devices. These technologies have transformed the way we watch television and movies, offering a more personalized and interactive viewing experience. With the ability to access a vast library of content on-demand, consumers can now easily discover and enjoy a diverse range of programming that caters to their unique interests and preferences.

The integration of infotainment systems into our vehicles has also significantly enhanced our driving experience. Modern cars are now equipped with sophisticated infotainment systems that offer a multitude of features, such as GPS navigation, hands-free calling, and voice-activated controls. These systems provide drivers and passengers with access to information and entertainment while on the road and contribute to increased safety and convenience.

Public spaces, too, have seen the integration of infotainment systems in the form of digital signage and interactive kiosks. These technologies provide valuable information and entertainment to people in various settings, such as shopping malls, airports, and train stations. By offering real-time updates on news, weather, and transportation schedules, as well as interactive maps and directories, digital signage and kiosks help to enhance the overall experience of navigating and utilizing public spaces.

The integration of infotainment systems into modern life has undoubtedly enriched our daily experiences and made information and entertainment more accessible than ever before. As technology continues to advance, we can expect to see even more innovative and immersive infotainment solutions that will further transform the way we live, work, and play.

The Future of Infotainment: Trends and Predictions

As we progress in the digital age, infotainment systems are poised to become even more advanced, versatile, and integrated into our daily lives. In this section, we will explore some of the most exciting trends and predictions for the future of infotainment, offering a glimpse into what we can expect in the coming years.

One of the most significant trends in the future of infotainment is the increasing convergence of various technologies. As devices become more interconnected, we can expect infotainment systems to seamlessly integrate with other aspects of our lives, such as home automation, wearable technology, and even our vehicles. This will lead to a more unified and personalized user experience, allowing individuals to easily access their preferred content and services across multiple platforms.

Another key trend in the future of infotainment is the rise of artificial intelligence (AI) and machine learning. These technologies have the potential to revolutionize the way we interact with infotainment systems, enabling them to become more intuitive, responsive, and tailored to individual preferences. For example, AI-powered voice assistants could become an integral part of infotainment systems, allowing users to control their devices through natural language commands and even engage in more complex conversations with their devices.

The increasing prevalence of augmented reality (AR) and virtual reality (VR) technologies also promises to have a significant impact on the future of infotainment. These immersive technologies have the potential to transform the way we consume content, offering new and engaging ways to experience entertainment, education, and communication. For instance, AR could be used to overlay digital information onto the physical world, while VR could transport users into entirely new environments and experiences.

In addition to these technological advancements, the future of infotainment will also be shaped by evolving consumer preferences and expectations. As audiences become more discerning and demanding, infotainment providers will need to adapt their offerings to cater to these changing tastes. This could include a greater emphasis on personalization, interactivity, and high-quality content, as well as a focus on addressing the needs of diverse and global audiences.

Finally, it is important to consider the potential ethical and societal implications of the future of infotainment. As these systems become more sophisticated and integrated into our lives, questions surrounding privacy, security, and the potential for manipulation will become increasingly relevant. It will be crucial for industry leaders, policymakers, and consumers alike to engage in ongoing discussions about the responsible development and use of infotainment technologies.

In conclusion, the future of infotainment is undoubtedly exciting, with numerous technological advancements and evolving consumer preferences set to shape the landscape in the coming years. As we continue to witness the ongoing evolution of infotainment systems, it is essential to remain mindful of the potential challenges and opportunities that lie ahead, ensuring that these innovations are harnessed for the benefit of all.

The Ongoing Evolution of Infotainment Systems

As we have explored throughout this chapter, the evolution of infotainment systems has been a fascinating journey, marked by continuous innovation and adaptation to the ever-changing landscape of technology and consumer preferences. From the early days of radio and television to the current era of digital technology and the internet, infotainment systems have come a long way in terms of their capabilities, reach, and impact on our daily lives.

The ongoing evolution of infotainment systems is driven by a myriad of factors, including advancements in technology, the increasing demand for personalized and immersive experiences, and the need to cater to a diverse and global audience. As we look ahead, it is clear that infotainment systems will continue to evolve, offering even more sophisticated and engaging content that seamlessly integrates into our lives.

One of the key trends shaping the future of infotainment is the increasing convergence of various media platforms, enabling users to access content across multiple devices and formats. This trend is already evident in the growing popularity of streaming services and the rise of multi-platform content providers. As technology continues to advance, we can expect to see even more seamless integration between different media platforms, allowing users to consume infotainment content in a truly personalized and convenient manner.

Another important trend is the growing importance of data and analytics in shaping the content and delivery of infotainment systems. As more and more data is generated by users and their devices, content providers will be able to leverage this information to create highly targeted and relevant experiences for their audiences. This will not only enhance the overall quality of infotainment content but also help providers to better understand and cater to the needs and preferences of their users.

Furthermore, the future of infotainment systems will likely be marked by an increasing emphasis on interactivity and user-generated content. As users become more accustomed to engaging with content in a participatory manner, infotainment providers will need to find innovative ways to incorporate user input and feedback into their offerings. This could take the form of interactive storytelling, collaborative content creation, or even the integration of gaming elements into traditional media formats.

Lastly, as the world becomes more interconnected and globalized, infotainment systems will need to adapt to cater to a diverse range of cultural and linguistic preferences. This will involve not only the creation of content that is relevant and appealing to different audiences but also the development of technology that enables seamless translation and localization of content across different languages and regions.

In conclusion, the ongoing evolution of infotainment systems is an exciting and dynamic process that promises to bring about even more engaging, immersive, and personalized experiences for users around the world. As technology continues to advance and our understanding of user preferences and behavior deepens, we can look forward to a future where infotainment systems are an integral and enriching part of our daily lives.

Chapter Summary

- Infotainment, a combination of information and entertainment, has evolved significantly over the years, with digital technology playing a pivotal role in shaping modern infotainment systems.
- Early innovations in infotainment can be traced back to newspaper comic strips, radio programs, and television shows that blended information and entertainment, with pioneers like Ted Turner and Oprah Winfrey leading the way.
- The rise of digital technology, including the internet, online platforms, and streaming services, has transformed the way we consume infotainment, making it more accessible and convenient.
- Social media platforms have become important sources of infotainment, allowing users to share and discover new content while providing a platform for content creators to reach wider audiences.
- Infotainment systems have integrated into various aspects of our lives, including smartphones, smart TVs, in-car systems,

and public spaces, enhancing our daily experiences and making information and entertainment more accessible.

- The future of infotainment will likely see increasing convergence of technologies, the rise of artificial intelligence and machine learning, and the growing prevalence of augmented and virtual reality, offering more immersive and engaging experiences.
- Evolving consumer preferences and expectations will shape the future of infotainment, with a greater emphasis on personalization, interactivity, and high-quality content catering to diverse and global audiences.
- As infotainment systems continue to evolve, it is crucial to consider the potential ethical and societal implications, such as privacy, security, and the potential for manipulation, and engage in ongoing discussions about responsible development and use of infotainment technologies.

2

COMPONENTS OF MODERN INFOTAINMENT SYSTEMS

A futuristic smart city, where various Internet of Things (IoT) devices seamlessly communicate with each other.

In today's fast-paced world, the automotive industry has evolved beyond merely providing transportation. Modern vehicles have become sophisticated machines that cater to the needs and desires of their occupants, offering a plethora of features and technologies aimed at enhancing the driving experience. One such innovation that has gained immense popularity in recent years is the infotainment system. As the name suggests, infotainment is a fusion of information and entertainment designed to keep drivers and passengers informed, entertained, and connected while on the road.

Infotainment systems have come a long way since their inception, transforming from basic radio and CD players to advanced, multi-functional systems that integrate various technologies and features. These systems have become an essential aspect of modern vehicles, with many consumers considering them a crucial factor when purchasing a new car. In this chapter, we will delve into the components of modern infotainment systems, exploring the role of connectivity, multimedia and entertainment features, navigation and location-based services, and the integration of voice control and artificial intelligence.

At the heart of every infotainment system is a central display, often a touchscreen, which serves as the primary interface for the user. This display allows drivers and passengers to access and control a wide range of functions, from audio and video playback to navigation and communication features. The user interface is designed to be intuitive and user-friendly, ensuring that even those who are not tech-savvy can easily navigate the system.

One of the most significant advancements in infotainment systems is the integration of smartphone connectivity. This feature allows users to seamlessly connect their smartphones to the vehicle's infotainment system, enabling them to access their favorite apps, make phone calls, send and receive messages, and even stream music and podcasts directly from their devices. This connectivity has revolutionized the way we interact with our vehicles, making it easier than ever to stay connected and entertained while on the road.

As we continue to explore the components of modern infotainment systems, it becomes evident that these systems are more than just a luxury or convenience feature. They have become integral to the driving experience, providing valuable information, entertainment, and connectivity options that enhance our time spent behind the wheel. In the following sections, we will delve deeper into the various aspects of infotainment systems, shedding light on the technologies and innovations that make them an essential component of today's vehicles.

The Role of Connectivity in Infotainment Systems

In today's fast-paced world, staying connected has become essential to our daily lives. As technology continues to evolve, so does the demand for seamless connectivity, even when we are on the move. This is where modern infotainment systems come into play. These advanced systems, found in most contemporary vehicles, offer many features that entertain, inform, and assist drivers and passengers alike. One of the most crucial components of these systems is connectivity, which serves as the backbone for a multitude of functions and features.

Connectivity is the lifeblood of modern infotainment systems, as it enables the exchange of data between the vehicle, its occupants, and the outside world. This constant flow of information allows for real-time updates, personalized content, and enhanced communication capabilities. In essence, connectivity transforms the vehicle into an extension of our digital lives, ensuring we remain connected to the people and information that matter most to us.

There are several ways in which infotainment systems can establish and maintain connectivity These include the following:

- **Cellular Connectivity:** This is the most common form of connectivity in infotainment systems. By leveraging existing cellular networks, vehicles can access the internet, enabling features such as real-time traffic updates, streaming music, and hands-free calling.
- **Wi-Fi Connectivity:** Some infotainment systems also offer built-in Wi-Fi capabilities, allowing passengers to connect

their devices to the internet without relying on their cellular data plans. This can be particularly useful for long road trips, where passengers may want to stream movies or play online games.

- **Vehicle-to-Everything (V2X) Communication:** This emerging technology enables vehicles to communicate with other vehicles, infrastructure, and even pedestrians, paving the way for a more connected and safer driving experience. V2X communication can provide real-time information on traffic conditions and road hazards, and even facilitate cooperative driving maneuvers.

Connectivity enables a wide range of services and applications that enhance the driving experience. Some of the most popular connected services include the following:

- **Real-Time Traffic Information:** By accessing up-to-date traffic data, infotainment systems can provide drivers with the most efficient routes, helping to save time and reduce fuel consumption.
- **Streaming Media:** Connectivity allows passengers to access their favorite music, podcasts, and audiobooks on-demand, ensuring they always have something to keep them entertained during their journey.
- **Social Media Integration:** Some infotainment systems offer integration with popular social media platforms, allowing drivers and passengers to stay connected with friends and family, share their location, or even post updates about their trip.
- **Remote Vehicle Services:** Connectivity enables remote access to various vehicle functions, such as starting the engine, adjusting climate controls, or locking and unlocking doors, all from a smartphone app.

In addition to entertainment and communication, connectivity is vital in enhancing vehicle safety and convenience. Connected services such as emergency assistance, roadside assistance, and vehicle diagnostics can provide drivers with peace of mind, knowing that help is just a button press away. Furthermore, connectivity enables the integration of advanced driver assistance systems (ADAS), which use data from various sensors and cameras to assist drivers in avoiding accidents and navigating complex driving situations.

In conclusion, connectivity is an indispensable component of modern infotainment systems, providing drivers and passengers with a wealth of information, entertainment, and convenience features. As technology continues to advance, we can expect even greater levels of connectivity and integration, paving the way for a more connected, informed, and enjoyable driving experience.

Multimedia and Entertainment Features

In today's fast-paced world, the demand for instant access to information and entertainment has never been greater. Modern infotainment systems have risen to the challenge, offering a plethora of multimedia and entertainment features that cater to the diverse needs and preferences of users. This section delves into the various multimedia and entertainment features that have become integral components of modern infotainment systems, enhancing the overall user experience and transforming the way we consume content.

One of the most fundamental aspects of any infotainment system is its ability to provide users with a rich audio and video experience. Modern systems support a wide range of formats, allowing users to enjoy their favorite music, podcasts, and videos with ease. In addition, high-quality speakers and advanced audio processing technologies ensure that users have an immersive and enjoyable listening experience, whether they are streaming content from their smartphones or playing it directly from the infotainment system's built-in storage.

The rise of streaming services has revolutionized the way we consume multimedia content. Modern infotainment systems have embraced this trend by integrating popular streaming services such as

Spotify, Apple Music, and Netflix, allowing users to access their favorite content on the go. This seamless integration offers users a vast library of music, movies, and TV shows and ensures they can enjoy personalized recommendations and curated playlists, further enhancing their entertainment experience.

In an increasingly connected world, social media and messaging platforms play a crucial role in our daily lives. Modern infotainment systems recognize this and offer integration with popular platforms such as Facebook, Twitter, and WhatsApp. This allows users to stay connected with their friends and family, receive real-time updates, and even share their experiences on the go, all without having to reach for their smartphones.

As technology continues to evolve, infotainment systems have started to incorporate gaming and interactive content, offering users a more engaging and immersive experience. From simple puzzle games to advanced virtual reality experiences, modern systems are equipped with the necessary hardware and software to support a wide range of gaming options. This helps keep passengers entertained during long journeys and opens up new possibilities for interactive learning and exploration.

A key aspect of modern infotainment systems is their ability to offer a high degree of customization and personalization. Users can tailor their systems to suit their preferences, from choosing the layout of the home screen to selecting their favorite color schemes and themes. Furthermore, advanced algorithms and artificial intelligence can analyze user behavior and preferences to offer personalized content recommendations, ensuring that the infotainment system remains relevant and engaging for each individual user.

In conclusion, multimedia and entertainment features have become essential components of modern infotainment systems, offering users a diverse and engaging experience that caters to their unique needs and preferences. As technology continues to advance, we can expect to see even more innovative and immersive features being integrated into infotainment systems, further transforming the way we consume content and stay connected on the go.

Navigation and Location-Based Services

In the era of smart devices and advanced technology, navigation and location-based services have become integral to modern infotainment systems. These services not only provide convenience and safety to drivers and passengers but also enhance the overall driving experience. In this section, we will delve into the various aspects of navigation and location-based services, their importance in infotainment systems, and the innovations that are shaping their future.

The primary purpose of navigation systems in infotainment is to provide accurate and real-time directions to drivers, ensuring that they reach their destinations efficiently and safely. With the integration of location-based services, these systems can offer personalized and context-aware information, such as nearby points of interest, traffic updates, and weather forecasts. This combination of navigation and location-based services has become a vital component of infotainment systems, as it significantly contributes to a seamless and enjoyable driving experience.

The backbone of navigation and location-based services is the Global Positioning System (GPS), which uses satellite signals to determine a vehicle's precise location. Modern infotainment systems employ advanced GPS receivers and mapping technologies to provide accurate and up-to-date maps, turn-by-turn directions, and real-time traffic information. Some systems even offer 3D mapping and augmented reality features, enhancing the visual representation of the route and making it easier for drivers to navigate unfamiliar roads.

The rise of smartphones and mobile applications has led to a significant shift in the way people access navigation and location-based services. Many drivers now prefer to use their smartphones for navigation, as they offer a familiar interface and access to popular third-party applications like Google Maps and Waze. To cater to this preference, infotainment systems have started to incorporate smartphone integration through technologies like Apple CarPlay and Android Auto, allowing users to access their preferred navigation apps and services directly from the infotainment system's display.

Another crucial aspect of navigation and location-based services is their integration with Advanced Driver Assistance Systems (ADAS). These systems use data from various sensors and cameras to provide features like adaptive cruise control, lane departure warnings, and parking assistance. By combining this data with navigation information, infotainment systems can offer route guidance that considers real-time traffic conditions, road hazards, and other factors that may affect the driving experience.

As technology continues to evolve, navigation and location-based services in infotainment systems are expected to become even more advanced and personalized. Innovations like 5G connectivity, machine learning, and the Internet of Things (IoT) will enable these systems to offer real-time updates, personalized recommendations, and seamless integration with other connected devices. Furthermore, the development of autonomous vehicles will likely lead to a significant transformation in the way navigation and location-based services are designed and utilized, as drivers will no longer need to focus on the road and can instead engage with the infotainment system in new and exciting ways.

In conclusion, navigation and location-based services play a crucial role in modern infotainment systems, providing drivers with accurate directions, real-time information, and personalized experiences. As technology continues to advance, we can expect these services to become even more sophisticated, further enhancing the driving experience and shaping the future of infotainment systems.

Integration of Voice Control and Artificial Intelligence

The integration of voice control and artificial intelligence (AI) has revolutionized the way we interact with infotainment systems, making them more user-friendly, efficient, and personalized. This section will delve into the importance of voice control and AI in modern infotainment systems, exploring their benefits and the potential for future advancements.

Voice control has become an essential feature in infotainment systems, allowing users to operate various functions without the need for manual input. This hands-free approach enhances user experience and contributes to safer driving by minimizing distractions. Voice control technology has evolved significantly over the years, with systems now able to understand and process complex commands, accents, and even different languages.

The integration of AI into voice control systems has further enhanced their capabilities. AI-powered voice assistants, such as Apple's Siri, Amazon's Alexa, and Google Assistant, have become increasingly popular in infotainment systems. These intelligent assistants can process natural language, learn from user interactions, and provide personalized recommendations based on user preferences and habits.

One of the most significant benefits of AI-powered voice control is its ability to adapt and improve over time. As users interact with the system, the AI learns from their preferences, speech patterns, and habits, allowing it to provide more accurate and relevant responses. This personalized approach ensures that the infotainment system remains engaging and useful throughout the life of the vehicle.

In addition to voice control, AI plays a crucial role in other aspects of infotainment systems. For example, AI algorithms can analyze user preferences and habits to curate personalized playlists, suggest new content, or even adjust the vehicle's climate control settings based on the user's comfort preferences. AI can also be used to optimize navigation routes based on real-time traffic data, weather conditions, and user preferences, ensuring a more efficient and enjoyable driving experience.

Looking ahead, the integration of voice control and AI in infotainment systems is expected to continue evolving. Advancements in natural language processing and machine learning will enable voice assistants to understand and process even more complex commands, making the user experience seamless and intuitive. Furthermore, as AI technology becomes more sophisticated, we can expect to see even greater levels of personalization and predictive capabilities in infotainment systems.

In conclusion, the integration of voice control and artificial intelligence has significantly enhanced the capabilities and user experience of modern infotainment systems. By allowing for hands-free operation, personalized recommendations, and adaptive learning, these technologies have made infotainment systems more engaging and useful than ever before. As voice control and AI technology continue to advance, we can expect to see even more innovative and exciting developments in the world of infotainment systems.

Chapter Summary

- Modern infotainment systems have evolved from basic radio and CD players to advanced, multi-functional systems that integrate various technologies and features, enhancing the driving experience.
- Connectivity is a crucial component of infotainment systems, enabling the exchange of data between the vehicle, its occupants, and the outside world, and providing real-time updates, personalized content, and enhanced communication capabilities.
- Multimedia and entertainment features have become essential components of modern infotainment systems, offering users a diverse and engaging experience that caters to their unique needs and preferences.
- Navigation and location-based services play a crucial role in modern infotainment systems, providing drivers with accurate directions, real-time information, and personalized experiences.
- The integration of voice control and artificial intelligence (AI) has revolutionized the way we interact with infotainment systems, making them more user-friendly, efficient, and personalized.
- AI-powered voice assistants, such as Apple's Siri, Amazon's Alexa, and Google Assistant, have become increasingly

popular in infotainment systems, providing personalized recommendations based on user preferences and habits.

- As concerns about data privacy and security continue to grow, future infotainment systems will likely place a greater emphasis on protecting users' information and developing more secure communication protocols and features that allow users to control the data collected and shared by their infotainment system.

3

INFOTAINMENT IN AUTOMOTIVE INDUSTRY

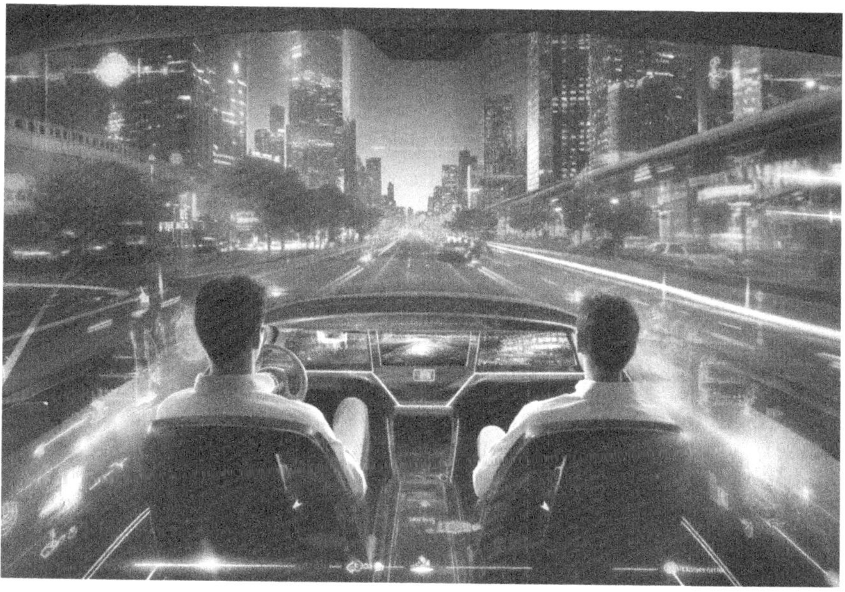

A futuristic infotainment system seamlessly integrated into the dashboard of a sleek, self-driving car.

Infotainment has become an essential aspect of the modern automotive industry. With rapid technological advancements and the increasing demand for connectivity and convenience, infotainment systems have evolved from simple radio receivers to sophisticated, multi-functional platforms that integrate navigation, communication, and entertainment features. These systems have transformed the way drivers and passengers interact with their vehicles, enhancing the overall driving experience and providing a seamless connection between the car and the digital world.

The automotive infotainment market has grown exponentially in recent years, driven by consumer demand for more advanced and user-friendly systems. As a result, car manufacturers and technology companies have been investing heavily in research and development to create innovative solutions that cater to their customers' ever-changing needs and preferences. This has led to the emergence of a highly competitive landscape, with industry players striving to differentiate themselves by offering unique and cutting-edge features that enhance the overall in-car experience.

In this chapter, we will delve into the fascinating world of automotive infotainment, exploring its evolution and development, key features and technologies, and the impact it has on driver experience and safety. We will also discuss future trends and innovations that are set to shape the future of infotainment in the automotive industry, highlighting the growing significance of this market segment in the broader context of the automotive ecosystem.

As we embark on this journey, it is important to recognize that infotainment systems are not just about providing entertainment and convenience; they also play a crucial role in improving driver safety and promoting responsible driving behavior. By offering real-time information, alerts, and assistance, these systems can help drivers make better decisions on the road, ultimately contributing to a safer and more enjoyable driving experience for all.

Evolution and Development of Automotive Infotainment Systems

The automotive infotainment industry has come a long way since its inception. From the early days of simple AM/FM radio systems to the sophisticated, interconnected systems of today, the evolution of automotive infotainment has been driven by technological advancements and the ever-changing demands of consumers. This section will explore the key milestones and developments that have shaped the landscape of automotive infotainment systems.

The journey of automotive infotainment began in the 1930s with the introduction of the first car radio. This simple AM radio system, developed by Motorola, marked the beginning of in-car entertainment and paved the way for future innovations. As technology progressed, FM radio was introduced in the 1950s, offering better sound quality and a wider range of stations for drivers to enjoy.

The 1960s saw the introduction of the 8-track tape player, which allowed drivers to listen to their favorite music on the go. This was followed by the cassette tape player in the 1970s, which quickly became the preferred choice for in-car audio due to its compact size and ease of use. The 1980s brought about the first in-car CD players, offering even better sound quality and the ability to store more music on a single disc.

As the 1990s rolled in, the focus shifted from audio entertainment to a more comprehensive infotainment experience. The introduction of GPS navigation systems revolutionized the way drivers navigated, providing real-time directions and maps. This decade also saw the rise of in-car television and DVD players, allowing passengers to watch movies and TV shows during long journeys.

The new millennium brought about a significant shift in the automotive infotainment landscape, with the advent of digital technology and the rise of the internet. The integration of Bluetooth technology enabled hands-free calling and wireless music streaming, while USB ports allowed for seamless connectivity with personal devices. The introduction of satellite radio provided drivers with an even wider range of entertainment options, including news, sports, and weather updates.

In recent years, the development of advanced infotainment systems has been heavily influenced by the proliferation of smartphones and mobile applications. Modern infotainment systems now offer touch-screen interfaces, voice recognition, and app integration, allowing drivers to access their favorite music, navigation, and communication apps directly from their vehicle's dashboard. These systems also support Apple CarPlay and Android Auto, further enhancing the connectivity between smartphones and in-car infotainment.

As we look toward the future, the evolution of automotive infotainment systems is expected to continue rapidly. With the advent of autonomous vehicles and the increasing importance of connectivity, we can expect to see even more advanced features and technologies integrated into our cars. From augmented reality navigation to AI-powered voice assistants, the possibilities for automotive infotainment are virtually limitless.

In conclusion, the evolution and development of automotive infotainment systems have been marked by significant technological advancements and a constant drive to meet the ever-changing demands of consumers. As we move forward, it is clear that infotainment will continue to play a crucial role in shaping the future of the automotive industry, enhancing the driving experience and offering new levels of connectivity and entertainment.

Key Features and Technologies in Modern Infotainment Systems

In today's fast-paced world, the automotive industry has evolved to meet the ever-changing needs and preferences of consumers. One such development is the integration of infotainment systems in vehicles, which has transformed the driving experience by providing a seamless blend of entertainment, information, and connectivity. This section delves into the key features and technologies that make modern infotainment systems indispensable to the contemporary driving experience.

Gone are the days of fumbling with buttons and knobs to control your car's entertainment system. Today's infotainment systems are equipped with high-resolution touchscreen displays allowing drivers and passengers to access various features with a tap or swipe. These displays come in various sizes and offer intuitive interfaces, making it easy for users to navigate through menus, adjust settings, and access information.

Smartphone integration is a crucial aspect of modern infotainment systems. Technologies such as Apple CarPlay and Android Auto enable drivers to connect their smartphones to the vehicle's infotainment system, allowing them to access their favorite apps, make phone calls, send messages, and stream music, all through the car's touchscreen display or voice commands. This integration enhances the driving experience and promotes safety by minimizing distractions.

Voice recognition technology has come a long way, and its integration into infotainment systems has made it easier for drivers to control various features without taking their hands off the wheel. By simply speaking commands, drivers can adjust the volume, change radio stations, make phone calls, and even input navigation destinations. This hands-free control greatly contributes to a safer and more enjoyable driving experience.

Built-in navigation systems have become a standard feature in many modern infotainment systems. These systems offer turn-by-turn directions, real-time traffic updates, and points of interest, ensuring that drivers can reach their destinations efficiently and safely. Some advanced navigation systems also incorporate augmented reality, displaying directional arrows and other information directly onto the windshield, further enhancing the user experience.

Connectivity is at the heart of modern infotainment systems, with features such as Wi-Fi hotspots, Bluetooth, and USB ports allowing drivers and passengers to stay connected on the go. These features enable users to access the internet, stream music, and connect multiple devices simultaneously, ensuring that everyone in the vehicle can enjoy a personalized and connected driving experience.

As technology continues to advance, infotainment systems need to keep up with the latest features and improvements. Over-the-Air (OTA) updates allow manufacturers to remotely update a vehicle's infotainment system, ensuring that drivers always have access to the most up-to-date software and features. This enhances the user experience and reduces the need for time-consuming and costly visits to dealerships for software updates.

In conclusion, modern infotainment systems have come a long way from their humble beginnings, offering a plethora of features and technologies that have revolutionized the driving experience. From touchscreen displays and smartphone integration to voice recognition and advanced navigation systems, these innovations have made driving more enjoyable, convenient, and safe. As technology continues to evolve, we can expect even more exciting developments in the world of automotive infotainment.

Impact of Infotainment on Driver Experience and Safety

The integration of infotainment systems in the automotive industry has significantly transformed the driving experience for motorists worldwide. These systems have enhanced the entertainment and connectivity aspects of driving and played a crucial role in improving driver safety. In this section, we will delve into the impact of infotainment on driver experience and safety, highlighting the benefits and potential challenges posed by these systems.

One of the most notable benefits of infotainment systems is the increased convenience and accessibility they offer to drivers. Modern infotainment systems provide a centralized hub for various functions like navigation, communication, and multimedia playback. This consolidation of features allows drivers to access essential information and perform tasks without having to fiddle with multiple devices or controls, thereby reducing distractions and enhancing overall safety.

Moreover, infotainment systems have significantly improved the quality of in-car entertainment. With features such as smartphone integration, streaming services, and advanced audio systems, drivers and passengers can now enjoy a wide range of entertainment options during their journeys. This makes long drives more enjoyable and helps alleviate driver fatigue, which is a significant contributor to road accidents.

Another critical aspect of infotainment systems is their role in promoting safer driving habits. Features such as voice recognition and hands-free controls enable drivers to perform tasks without taking their hands off the wheel or their eyes off the road. Additionally, advanced driver assistance systems (ADAS) integrated into infotainment systems provide real-time feedback on driving performance, alerting drivers to potential hazards and promoting safer driving behaviors.

However, it is essential to acknowledge the potential challenges and risks associated with infotainment systems. One such concern is the possibility of information overload, as drivers may be bombarded with too much data and notifications, leading to distractions and compromised safety. To mitigate this issue, automotive manufacturers must design intuitive and user-friendly interfaces that prioritize essential information and minimize distractions.

Another concern is the potential for over-reliance on infotainment systems and ADAS features, which may lead to complacency and reduced driver vigilance. Drivers need to understand that these systems are meant to assist and not replace their judgment and responsibility on the road.

In conclusion, the integration of infotainment systems in the automotive industry has significantly impacted driver experience and safety. While these systems offer numerous benefits, such as increased convenience, enhanced entertainment, and improved driving habits, it is crucial to address potential challenges and risks to ensure that infotainment systems continue to contribute positively to driver safety. As technology evolves, it is essential for automotive manufacturers and drivers alike to strike a balance between leveraging the benefits of infotainment systems and maintaining the necessary vigilance and responsibility on the road.

Future Trends and Innovations in Automotive Infotainment

As we venture further into the 21st century, the automotive industry continues to evolve rapidly, with infotainment systems playing an increasingly significant role in the overall driving experience. In this section, we will explore some of the most exciting and innovative trends that are set to shape the future of automotive infotainment.

One of the most promising developments in automotive infotainment is the integration of AI-powered voice assistants. These intelligent systems are designed to understand and respond to natural language commands, allowing drivers to interact with their vehicles in a more intuitive and hands-free manner. This not only enhances the user experience but also helps to minimize distractions and improve overall safety. As AI technology continues to advance, we can expect voice assistants to become even more sophisticated, capable of understanding complex requests and providing personalized recommendations based on user preferences and driving habits.

Another exciting innovation in the realm of automotive infotainment is the use of augmented reality and heads-up displays. These technologies project relevant information, such as navigation instructions, vehicle speed, and safety alerts, directly onto the windshield, allowing drivers to access critical data without taking their eyes off the road. As AR technology continues to mature, we can expect to see even more immersive and contextually-aware displays that seamlessly integrate with the surrounding environment, providing drivers with a truly futuristic driving experience.

The future of automotive infotainment also lies in the realm of vehicle-to-everything (V2X) communication. This technology enables vehicles to communicate with other vehicles, infrastructure, and even pedestrians, providing real-time information on traffic conditions, road hazards, and other relevant data. By leveraging V2X communication, infotainment systems can offer drivers more accurate and timely information, ultimately enhancing safety and efficiency on the road. As more vehicles become equipped with V2X capabilities, we expect to see a more connected and intelligent transportation ecosystem emerge.

As infotainment systems become more advanced, they will also offer greater levels of personalization and customization. This means that drivers can tailor their infotainment experience to their specific preferences, from the interface layout to the types of content and services available. Additionally, infotainment systems will be able to learn from user behavior and adapt accordingly, providing a more seamless and enjoyable experience for drivers and passengers alike.

Finally, the future of automotive infotainment will see greater integration with smart homes and Internet of Things (IoT) devices. This will enable drivers to control various aspects of their home environment, such as lighting, temperature, and security systems, directly from their vehicle's infotainment system. This connectivity level will provide added convenience and help create a more unified and cohesive user experience across multiple devices and platforms.

In conclusion, the future of automotive infotainment is bright, with numerous innovations and trends poised to revolutionize the way we interact with our vehicles. As technology continues to advance, we can expect to see even more sophisticated and immersive infotainment systems that enhance the driving experience while prioritizing safety and efficiency.

The Growing Significance of Infotainment in the Automotive Industry

In conclusion, the role of infotainment in the automotive industry has grown exponentially over the past few decades, transforming the way drivers and passengers interact with their vehicles. As technology continues to advance, infotainment systems have become an integral part of the driving experience, offering a wide range of features and functionalities that cater to the diverse needs and preferences of consumers.

The evolution of automotive infotainment systems has been driven by the rapid development of cutting-edge technologies, such as touchscreens, voice recognition, and smartphone integration. These advancements have enhanced the user experience and contributed to the overall safety and convenience of driving. By providing real-time traffic updates, navigation assistance, and hands-free communication, modern infotainment systems have significantly reduced the risk of driver distraction and improved situational awareness on the road.

Moreover, the growing significance of infotainment in the automotive industry can be attributed to the increasing consumer demand for seamless connectivity and personalized entertainment options. As a result, automakers are constantly striving to innovate and integrate new features into their infotainment systems, such as augmented reality, artificial intelligence, and advanced driver assistance systems (ADAS). These innovations are expected to further revolutionize the way we interact with our vehicles, making the driving experience more enjoyable, efficient, and safe.

In addition to enhancing the driving experience, the rise of infotainment systems has also created new opportunities for businesses within the automotive industry. The integration of various technologies and services has led to the emergence of new revenue streams, such as subscription-based content and data-driven services. Furthermore, the increasing importance of infotainment has also fostered collaboration between automakers, technology companies, and content providers, paving the way for new partnerships and business models.

As we look toward the future, it is evident that infotainment will continue to play a crucial role in shaping the automotive industry. With the advent of autonomous vehicles and the growing trend towards shared mobility, the role of infotainment systems is expected to evolve even further, providing passengers with a fully immersive and personalized experience. As a result, automakers and technology providers must continue to invest in research and development to stay ahead of the curve and meet the ever-changing needs of consumers.

In summary, the growing significance of infotainment in the automotive industry is a testament to the transformative power of technology and its ability to enhance our lives in countless ways. As we continue to witness the rapid evolution of infotainment systems, it is clear that they will remain an essential component of the driving experience, shaping the future of mobility and redefining our relationship with our vehicles.

Chapter Summary

- Infotainment systems have evolved from simple radio receivers to sophisticated, multi-functional platforms that integrate navigation, communication, and entertainment features, transforming the way drivers and passengers interact with their vehicles.
- The automotive infotainment market has grown exponentially due to consumer demand for advanced and user-friendly systems, leading to a highly competitive landscape with industry players striving to differentiate themselves through unique and cutting-edge features.
- The evolution of automotive infotainment systems has been driven by technological advancements, such as touchscreens, voice recognition, and smartphone integration, which have enhanced the user experience and contributed to overall safety and convenience.
- Modern infotainment systems offer many features and technologies, including touchscreen displays, smartphone integration, voice recognition and control, navigation systems, connectivity features, and over-the-air (OTA) updates.
- Infotainment systems have significantly impacted driver experience and safety by providing increased convenience, enhanced entertainment, and promoting safer driving habits through features such as voice recognition and hands-free controls.
- Future trends and innovations in automotive infotainment include the integration of artificial intelligence (AI) and voice assistants, augmented reality (AR) and heads-up displays (HUDs), vehicle-to-everything (V2X)

communication, personalization and customization, and integration with smart home and IoT devices.

- The growing significance of infotainment in the automotive industry has created new opportunities for businesses, fostering collaboration between automakers, technology companies, and content providers, and paving the way for new partnerships and business models.
- As the automotive industry continues to evolve, infotainment systems will remain an essential component of the driving experience, shaping the future of mobility and redefining our relationship with our vehicles.

4

USER INTERFACE DESIGN FOR INFOTAINMENT SYSTEMS

An infotainment system UI that takes inspiration from the vibrant colors and geometric shapes of the 1980s retro-futuristic aesthetic.

In today's fast-paced, technology-driven world, infotainment systems have become an integral part of our daily lives. From in-car entertainment and navigation systems to smart TVs and home automation devices, these systems provide a seamless blend of information and entertainment, making our lives more convenient, enjoyable, and connected. At the heart of these infotainment systems lies the user interface (UI), which serves as the bridge between the user and the technology. A well-designed UI is essential for ensuring that users can easily access and interact with the wealth of features and functionalities offered by these systems.

User interface design for infotainment systems is a specialized field that combines elements of graphic design, human-computer interaction, and cognitive psychology to create intuitive, visually appealing, and user-friendly interfaces. The primary goal of UI design in infotainment systems is to provide users with a seamless and enjoyable experience, allowing them to easily navigate the system, access relevant information, and control various features and functionalities with minimal effort.

As infotainment systems continue to evolve and become more sophisticated, the importance of effective UI design cannot be overstated. With an ever-growing array of features and functionalities, users are faced with an increasing amount of information and options, making it crucial for designers to create interfaces that are easy to understand, navigate, and interact with. Furthermore, as infotainment systems become more deeply integrated into our daily lives, the need for accessible and inclusive design becomes paramount, ensuring that users of all ages, abilities, and backgrounds can fully benefit from these technologies.

In this chapter, we will delve into the world of user interface design for infotainment systems, exploring the principles and best practices that guide the creation of effective, engaging, and accessible interfaces. We will discuss the delicate balance between aesthetics and functionality, as well as the importance of designing for accessibility and inclusivity. Through case studies of successful infotainment UI designs, we will examine the various elements and techniques that contribute to a seamless and enjoyable user experience. Finally, we will look ahead to the future of user interface design in infotainment systems, considering the emerging trends and technologies that will shape the way we interact with these systems in the years to come.

Principles of Effective Infotainment UI Design

In the rapidly evolving world of infotainment systems, user interface (UI) design plays a crucial role in ensuring that users can access and interact with content in a seamless and enjoyable manner. As infotainment systems become more sophisticated, designers must strike a balance between creating visually appealing interfaces and maintaining ease of use. This section will explore the fundamental principles of effective infotainment UI design, which can help designers create interfaces that look great and provide a satisfying user experience.

Clarity and simplicity: The primary goal of any UI design is to facilitate user interaction with the system. To achieve this, designers must prioritize clarity and simplicity in their designs. This means using clear, concise language, intuitive navigation, and easily recognizable icons and buttons. Designers can create a more streamlined and user-friendly interface by minimizing visual clutter and focusing on essential elements.

Consistency: Consistency is a vital aspect of UI design, as it helps users build a mental model of how the system works. By maintaining a consistent layout, color scheme, typography, and iconography across all screens and menus, designers can make it easier for users to navigate the interface and understand how different elements relate to one another.

Feedback and responsiveness: Infotainment systems must provide users with immediate feedback to confirm that their actions have been acknowledged and executed. This can be achieved through visual cues, such as highlighting a selected button or displaying a progress bar, as well as auditory cues, like a click or beep sound. Responsiveness is also crucial, as users expect the system to react quickly to their inputs. Delays in response can lead to frustration and a negative user experience.

Flexibility and customization: As infotainment systems cater to diverse users with varying preferences and needs, it is essential to offer flexibility and customization options. This can include adjustable font sizes, color themes, and the ability to rearrange or hide certain elements on the screen. Designers can enhance user satisfaction and engagement by providing users with the tools to tailor the interface to their liking.

Anticipating user needs: A well-designed infotainment UI should anticipate the user's needs and provide relevant information and options at the right time. This can be achieved through the use of context-sensitive menus, predictive text input, and smart recommendations based on user behavior and preferences. By proactively addressing user needs, designers can create a more efficient and enjoyable user experience.

Minimizing cognitive load: Infotainment systems often involve multitasking, such as navigating a map while listening to music or making a phone call. To prevent users from becoming overwhelmed, designers must minimize the cognitive load by presenting information in a clear, organized manner and avoiding information overload. This can be achieved through the use of hierarchical structures, grouping related elements together, and providing visual cues to guide users through the interface.

In conclusion, effective infotainment UI design is a delicate balance of aesthetics, functionality, and usability. By adhering to these principles, designers can create visually appealing interfaces and provide a satisfying and efficient user experience. As infotainment systems continue to evolve, it is crucial for designers to stay up-to-date with the latest trends and technologies to ensure that their designs remain relevant and engaging for users.

Balancing Aesthetics and Functionality in Infotainment Interfaces

In the world of infotainment systems, the user interface (UI) plays a crucial role in delivering a seamless and enjoyable experience for the end user. A well-designed UI enhances the system's overall aesthetics and ensures users can easily access and interact with the content. Striking the right balance between aesthetics and functionality is a critical aspect of UI design in infotainment systems, and this section will delve into the various considerations and best practices that designers should keep in mind while working on such projects.

First and foremost, it is essential to understand that aesthetics and functionality are not mutually exclusive. A visually appealing interface can significantly improve the user experience by creating a positive first impression and fostering a sense of trust and credibility. However, it is equally important to ensure that the interface is easy to navigate and use, as even the most visually stunning design can fall short if users struggle to find the information or features they need.

To achieve the perfect balance between aesthetics and functionality, designers should focus on the following fundamental principles:

Clarity and simplicity: The primary goal of any infotainment UI should be to make it easy for users to access and consume content. This can be achieved by using clear and concise language, intuitive navigation, and a clean layout that minimizes visual clutter. Designers should avoid using overly complex graphics or animations that may distract users from the main content.

Consistency: Maintaining a consistent visual language throughout the interface is crucial for creating a cohesive and professional look. This includes using a uniform color scheme, typography, and iconography that align with the overall brand identity. Consistency also extends to the layout and placement of various UI elements, ensuring that users can quickly familiarize themselves with the interface and navigate it with ease.

Responsiveness and adaptability: Infotainment systems are often accessed through a variety of devices, such as smartphones, tablets, and in-car displays. Designers must ensure the UI is responsive and adaptable to different screen sizes and resolutions, providing an optimal viewing experience for all users. This may involve creating multiple layouts or using flexible design elements that can automatically adjust to the available screen space.

User feedback: Providing users with clear and immediate feedback is essential for creating an engaging and interactive experience. This can be achieved through subtle visual cues, such as highlighting buttons when they are pressed or using animations to indicate that an action has been performed successfully. Feedback can also be provided through sound or haptic feedback, depending on the device and user preferences.

Accessibility: Designing for inclusivity is a crucial aspect of UI design, ensuring that infotainment systems can be used by individuals with varying abilities and needs. This may involve incorporating features such as adjustable font sizes, high-contrast color schemes, and alternative input methods, such as voice commands or gesture controls.

In conclusion, balancing aesthetics and functionality in infotainment interfaces is a delicate process that requires careful consideration of various design principles and user needs. By focusing on clarity, consistency, responsiveness, user feedback, and accessibility, designers can create visually appealing and highly functional infotainment systems that cater to a diverse audience and provide an enjoyable user experience.

Designing for Accessibility and Inclusivity in Infotainment Systems

In today's diverse and interconnected world, it is more important than ever to design infotainment systems that cater to the needs of all users, regardless of their age, abilities, or cultural background. Accessibility and inclusivity should be at the forefront of every user interface (UI) designer's mind when creating infotainment systems. This section will explore the importance of designing for accessibility and inclusivity, as well as provide practical tips and guidelines for achieving these goals in infotainment UI design.

The importance of accessibility and inclusivity in infotainment systems cannot be overstated. As the global population continues to age and the prevalence of disabilities increases, it is crucial that infotainment systems are designed to be usable by everyone, including those with visual, auditory, cognitive, or physical impairments. Furthermore, as the world becomes more interconnected, infotainment systems must be designed to accommodate users from various cultural backgrounds and language proficiencies. By designing for accessibility and inclusivity, UI designers can ensure that their infotainment systems are enjoyed by the widest possible audience and that no one is excluded from the benefits of these systems.

To design infotainment systems that are both accessible and inclusive, UI designers should adhere to the following guidelines:

Follow established accessibility standards: There are several established accessibility standards and guidelines, such as the Web Content Accessibility Guidelines (WCAG) and the ADA Standards for Accessible Design. These guidelines provide a wealth of information on how to design UI elements that are accessible to users with various disabilities.

Use clear and simple language: To accommodate users with varying language proficiencies and cognitive abilities, it is essential to use clear and simple language in all text elements of the infotainment system. This includes labels, instructions, and error messages. Additionally, providing translations or localization options can further enhance the inclusivity of the system.

Ensure high contrast and legibility: Users with visual impairments may struggle to read text or discern UI elements if there is insufficient contrast between the text and background colors. Designers should ensure that all text and UI elements have a high contrast ratio and are easily legible.

Provide alternative input methods: Some users may have difficulty using traditional input methods, such as touchscreens or buttons, due to physical impairments. Designers should consider incorporating alternative input methods, such as voice control or gesture recognition, to accommodate these users.

Design for various screen sizes and resolutions: Infotainment systems are used on a wide range of devices, from smartphones to large in-car displays. Designers should ensure their UI is adaptable and functional on various screen sizes and resolutions.

Offer customizable settings: Allowing users to customize the appearance and functionality of the infotainment system can greatly enhance its accessibility and inclusivity. This may include options for adjusting text size, color schemes, or audio settings.

By following these guidelines and prioritizing accessibility and inclusivity in the design process, UI designers can create infotainment systems that are enjoyable and usable by all users, regardless of their abilities or background. This not only benefits the users themselves but also contributes to a more inclusive and connected world.

Case Studies: Successful Infotainment UI Designs

This section will explore real-world examples of successful infotainment user interface (UI) designs. These case studies will provide valuable insights into the strategies and techniques employed by industry leaders to create engaging, user-friendly, and accessible infotainment systems.

Case Study 1: Apple CarPlay

Apple CarPlay is a prime example of a successful infotainment UI design. Launched in 2014, CarPlay allows users to connect their iPhones to their car's infotainment system, providing a seamless and familiar interface for accessing music, navigation, and communication apps.

One of the key strengths of CarPlay's UI design is its simplicity. The interface features large, easy-to-read icons and a minimalistic design that closely resembles the iPhone's home screen. This familiarity allows users to navigate the system with ease, reducing the learning curve and minimizing driver distraction.

Additionally, CarPlay's UI design prioritizes accessibility and inclusivity. The system supports Siri voice commands, enabling users with visual impairments or limited dexterity to interact with the infotainment system without needing to touch the screen. Furthermore, CarPlay is available in multiple languages, ensuring a broad range of users can enjoy its features.

Case Study 2: Tesla Model S Infotainment System

Tesla's Model S infotainment system is another shining example of successful UI design in the automotive industry. The system features a massive 17-inch touchscreen display that serves as the central hub for controlling various vehicle functions, including climate control, navigation, and entertainment.

The Model S infotainment system's UI design is characterized by its clean, modern aesthetic and intuitive layout. The interface is divided into two main sections: the top half displays navigation and media information, while the bottom half provides access to vehicle controls and settings. This layout enables users to quickly locate and access the desired functions with minimal distraction.

Tesla's infotainment system also incorporates several accessibility features, such as voice commands and customizable font sizes. These design choices ensure that a diverse range of users can comfortably interact with the system.

Case Study 3: Spotify for Android Auto

Spotify's integration with Android Auto demonstrates how a popular music streaming service can successfully adapt its UI design for an infotainment platform. The Android Auto version of Spotify features a simplified interface with large, easy-to-tap icons and limited on-screen options.

This streamlined design allows users to quickly access their favorite playlists, browse through recommended songs, and control playback without getting overwhelmed by excessive visual clutter. Moreover, Spotify for Android Auto supports voice commands, enabling users to search for songs, artists, and albums without taking their hands off the wheel.

In conclusion, these case studies showcase the importance of thoughtful UI design in creating successful infotainment systems. By prioritizing simplicity, familiarity, accessibility, and inclusivity, industry leaders like Apple, Tesla, and Spotify have developed infotainment interfaces that are visually appealing, highly functional, and user-friendly. As technology continues to evolve, we can expect to see even more innovative and engaging UI designs in the infotainment space.

The Future of User Interface Design in Infotainment Systems

As we have explored throughout this chapter, user interface design plays a crucial role in the development and success of infotainment systems. From adhering to the principles of effective UI design to ensuring accessibility and inclusivity, designers must continually adapt and innovate to meet the ever-evolving needs and expectations of users. As we look toward the future of user interface design in infotainment systems, several key trends and advancements are expected to shape the industry and redefine the way we interact with these systems.

One of the most significant trends in the future of UI design for infotainment systems is the increasing integration of artificial intelligence (AI) and machine learning. These technologies have the potential to revolutionize the way users interact with infotainment systems by enabling more personalized, context-aware, and adaptive experiences. For example, AI-driven voice assistants can provide a more natural and intuitive way for users to access information and control various functions within the system. Additionally, machine learning algorithms can analyze user behavior and preferences to tailor the interface and content recommendations, ensuring a more relevant and engaging experience.

Another key development in the future of infotainment UI design is the rise of augmented reality (AR) and virtual reality (VR) technologies. These immersive technologies can offer new and exciting ways for users to interact with infotainment systems, blurring the lines between the digital and physical worlds. For instance, AR-based interfaces can overlay digital information onto the user's real-world environment, providing contextually relevant information and enhancing the overall experience. Similarly, VR can transport users to entirely new environments, offering unique and engaging infotainment experiences that were previously unimaginable.

The increasing importance of data privacy and security will also play a significant role in shaping the future of user interface design for infotainment systems. As these systems collect and store vast amounts of user data, designers must prioritize the protection of this information and ensure that users have control over their data. This will likely involve the development of more transparent and user-friendly privacy settings, as well as the incorporation of advanced security measures to safeguard user data from potential threats.

Lastly, as the world becomes more interconnected, the future of infotainment UI design will likely involve greater integration with other devices and systems. This could include seamless connectivity with smartphones, wearables, and smart home devices, allowing users to access and control their infotainment systems from various touchpoints. This increased connectivity will require designers to create consistent and intuitive user experiences across multiple platforms, ensuring that users can easily navigate and interact with their infotainment systems, regardless of the device they are using.

In conclusion, the future of user interface design in infotainment systems is poised for significant advancements and transformations. By embracing emerging technologies, prioritizing data privacy and security, and ensuring seamless integration with other devices, designers can create innovative and engaging infotainment experiences that cater to the diverse needs and preferences of users. As we move forward, designers need to remain adaptable and open to change as the world of infotainment continues to evolve at a rapid pace.

Chapter Summary

- User interface (UI) design plays a crucial role in infotainment systems, ensuring users can easily access and interact with the wealth of features and functionalities offered by these systems.
- Effective infotainment UI design involves balancing aesthetics and functionality, prioritizing clarity, simplicity, consistency, feedback, responsiveness, flexibility, and anticipating user needs.
- Designers should strive to create visually appealing interfaces that are also user-friendly, easy to navigate, and provide a satisfying user experience.
- Accessibility and inclusivity are essential aspects of UI design, ensuring that infotainment systems cater to users of all ages, abilities, and backgrounds.
- Following established accessibility standards, using clear language, ensuring high contrast and legibility, providing alternative input methods, and designing for various screen sizes and resolutions are key guidelines for creating accessible and inclusive infotainment systems.
- Successful infotainment UI designs, such as Apple CarPlay, Tesla Model S Infotainment System, and Spotify for Android Auto, prioritize simplicity, familiarity, accessibility, and inclusivity to create engaging and user-friendly experiences.
- The future of UI design in infotainment systems will likely involve the integration of artificial intelligence, augmented reality, virtual reality, increased data privacy and security, and seamless connectivity with other devices and systems.
- Designers must remain adaptable and open to change as the world of infotainment continues to evolve rapidly, embracing emerging technologies and prioritizing user needs and preferences.

5

CONNECTIVITY AND INTEGRATION IN INFOTAINMENT

A futuristic smart city, where various Internet of Things (IoT) devices seamlessly communicate with each other.

The world of infotainment has come a long way since its inception. From the early days of radio broadcasts and black-and-white television sets to the current era of streaming services, smart devices, and the Internet of Things (IoT), the way we consume information and entertainment has undergone a massive transformation. At the heart of this evolution lies the ever-growing importance of connectivity and integration in infotainment systems.

Connectivity, in the context of infotainment, refers to the ability of various devices, platforms, and services to communicate and share information with one another. Integration, on the other hand, is the seamless blending of these connected elements to create a unified and user-friendly experience. Together, these two aspects have revolutionized the way we access, consume, and interact with information and entertainment content.

The journey of connectivity in infotainment began with the introduction of wired connections, such as radio and television broadcasts, which allowed users to access a limited range of content through dedicated devices. As technology advanced, we witnessed the emergence of cable and satellite television, which expanded the variety of content available to viewers and enabled the integration of additional services, such as on-demand programming and interactive features.

The advent of the internet marked a significant turning point in the evolution of connectivity and integration in infotainment. With the World Wide Web came the possibility of accessing an almost limitless array of content from around the globe and the ability to share and interact with that content in real time. This development paved the way for the rise of streaming services, social media platforms, and other digital channels that have since become integral components of our infotainment landscape.

In recent years, the proliferation of smart devices and IoT technology has further accelerated the growth of connectivity and integration in infotainment. Today, we can access our favorite content and services through many devices, such as smartphones, tablets, smart TVs, and even voice-activated assistants. This has led to a more personalized and immersive infotainment experience, as users can now tailor their content consumption to their preferences and enjoy it across multiple platforms.

As we look ahead, the future of connectivity and integration in infotainment promises to be even more exciting and dynamic. With the impending rollout of 5G networks and the continued development of cutting-edge technologies, we can expect to see even greater levels of interconnectivity and seamless integration between devices, platforms, and services.

In this chapter, we will delve deeper into the various facets of connectivity and integration in infotainment, exploring the role of smart devices and IoT, the importance of cross-platform compatibility, the impact of 5G and future technologies, and the challenges and opportunities that lie ahead. By understanding these aspects, we can better appreciate the transformative power of connectivity and integration in shaping the future of infotainment.

The Role of Smart Devices and IoT in Infotainment Integration

The rapid proliferation of smart devices and the Internet of Things (IoT) has revolutionized the way we consume and interact with information and entertainment. In this section, we will explore the role of these technologies in infotainment integration and how they have transformed the landscape of this industry.

Smart devices like smartphones, tablets, and smart TVs have become integral to our daily lives. These devices provide us with communication and productivity tools and serve as gateways to a vast array of infotainment content. From streaming music and video to accessing news and social media, smart devices have made it easier than ever to stay informed and entertained on the go.

The integration of infotainment into smart devices has led to the development of various applications and platforms that cater to users' diverse interests and preferences. For instance, streaming services like Netflix, Spotify, and YouTube have become household names, offering a plethora of content that can be accessed anytime, anywhere. Furthermore, the rise of voice-activated virtual assistants, such as Amazon's Alexa, Google Assistant, and Apple's Siri, has enabled users to access infotainment content through simple voice commands, making the experience even more seamless and convenient.

The Internet of Things (IoT) refers to the network of interconnected devices and objects that communicate and exchange data with each other. In the context of infotainment, IoT plays a crucial role in connecting various devices and platforms, enabling a more integrated and immersive experience for users.

One of the most prominent examples of IoT in infotainment is the connected car. Modern vehicles are equipped with advanced infotainment systems that integrate navigation, music, and communication features, all accessible through a central interface. By connecting to the internet and other devices, such as smartphones and wearables, these systems can provide real-time traffic updates, personalized music playlists, and hands-free communication options, enhancing the overall driving experience.

Similarly, IoT has also made its way into our homes, with smart speakers, smart TVs, and other connected appliances working together to create a seamless infotainment ecosystem. Users can now control their entertainment systems, receive news updates, and even manage their home environment through a single, unified platform.

The combination of smart devices and IoT has created a powerful synergy that has propelled the infotainment industry to new heights. This fusion has enabled users to access and control their infotainment content across multiple devices, creating a seamless and personalized experience.

For instance, a user can start watching a movie on their smart TV, pause it, and then resume watching it on their tablet or smartphone without missing a beat. Similarly, they can listen to their favorite podcast on their morning commute and then continue listening on their smart speaker when they arrive home. This level of integration and convenience has become the new norm in the infotainment industry, and it is only expected to grow as technology continues to advance.

In conclusion, the role of smart devices and IoT in infotainment integration cannot be overstated. These technologies have transformed the way we consume and interact with information and entertainment, making it more accessible, personalized, and immersive than ever before. As we look to the future, it is clear that the synergy between smart devices and IoT will continue to shape the infotainment landscape, offering exciting new possibilities and experiences for users around the world.

Cross-Platform Compatibility: Bridging the Gap Between Devices and Services

As infotainment systems continue to evolve, one of the most significant challenges developers and consumers face is ensuring cross-platform compatibility. This means creating a seamless experience for users, regardless of the device or service they are using. In this section, we will explore the importance of cross-platform compatibility in infotainment, the steps being taken to bridge the gap between devices and services, and the benefits it brings to both users and the industry as a whole.

Cross-platform compatibility is essential in today's interconnected world, where users expect to access their favorite content and services on any device at any time. This expectation extends to infotainment systems, which must be able to communicate with a wide range of devices, from smartphones and tablets to wearables and smart home appliances. By ensuring compatibility across platforms, infotainment systems can offer users a consistent and enjoyable experience, regardless of the device they are using.

One of the key challenges in achieving cross-platform compatibility is the sheer number of devices and services available on the market. With each device and service utilizing its own unique software, hardware, and communication protocols, creating a seamless experience can be a daunting task. To address this issue, industry leaders have begun to develop standardized frameworks and protocols that allow for easier integration between devices and services.

For example, the Automotive Grade Linux (AGL) project is an open-source initiative that aims to create a unified platform for automotive infotainment systems. By providing a common framework for developers, AGL helps to streamline the process of integrating various devices and services, ultimately leading to a more cohesive user experience.

Another approach to cross-platform compatibility is the use of application programming interfaces (APIs) and software development kits (SDKs). These tools allow developers to create applications and services that can easily communicate with one another, regardless of the underlying platform. By utilizing APIs and SDKs, infotainment systems can offer a wide range of features and services, from navigation and voice recognition to streaming music and social media integration.

The benefits of cross-platform compatibility in infotainment are numerous. For users, it means a more seamless and enjoyable experience, as they can access their favorite content and services on any device without worrying about compatibility issues. For developers, it simplifies the process of creating new applications and services, as they can build on existing frameworks and protocols. And for the industry as a whole, it fosters innovation and collaboration, as companies can work together to create new and exciting infotainment experiences.

In conclusion, cross-platform compatibility is a crucial aspect of modern infotainment systems, as it allows for seamless integration between devices and services. By embracing standardized frameworks, APIs, and SDKs, the industry can overcome the challenges associated with compatibility and deliver a more cohesive and enjoyable experience for users. As technology continues to advance, we can expect to see even greater levels of connectivity and integration in infotainment, ultimately leading to a more interconnected and immersive experience for all.

The Impact of 5G and Future Technologies on Infotainment Connectivity

The advent of 5G technology has the potential to revolutionize the infotainment industry by providing lightning-fast connectivity and seamless integration of various devices and services. As we move towards a world where everything is connected, the impact of 5G and other emerging technologies on infotainment cannot be overstated. In this section, we will explore how these advancements are shaping the future of infotainment connectivity and what it means for consumers and businesses alike.

One of the most significant advantages of 5G technology is its ability to offer unparalleled speed and bandwidth compared to its predecessors. With data transfer rates up to 100 times faster than 4G, 5G enables the streaming of high-quality content, such as 4K and 8K videos, with virtually no buffering or lag. This improvement in speed and bandwidth allows for a more immersive and interactive infotainment experience, as users can now access and consume content in real time without disruptions.

The increased speed and bandwidth of 5G technology also contribute to an enhanced user experience in infotainment. With faster connectivity, users can access a wider range of content and services, such as augmented reality (AR) and virtual reality (VR) applications, which were previously limited by the constraints of 4G networks. Furthermore, 5G enables the seamless integration of multiple devices, allowing users to switch between their smartphones, tablets, and other smart devices without any loss in quality or performance.

Another significant impact of 5G on infotainment connectivity is the rise of edge computing. By processing data closer to the source, edge computing reduces latency and improves the overall performance of infotainment systems. This technology allows for real-time data analysis and decision-making, which is crucial for applications such as autonomous vehicles and smart city infrastructure. As a result, infotainment systems can now provide users with more accurate and up-to-date information, enhancing their overall experience.

The rapid advancements in 5G and other emerging technologies have paved the way for new business models and opportunities within the infotainment industry. For instance, the increased speed and bandwidth provided by 5G networks enable content creators to offer more immersive and interactive experiences, such as live events and virtual concerts. Additionally, businesses can now leverage the power of 5G to develop innovative solutions and services, such as personalized advertising and location-based content delivery, which were previously not feasible due to the limitations of 4G networks.

As the infotainment industry continues to evolve, it is crucial for businesses and consumers alike to stay informed about the latest developments in connectivity and integration technologies. By embracing the potential of 5G and other emerging technologies, we can unlock new possibilities and experiences in infotainment, ultimately transforming how we consume and interact with content.

In conclusion, the impact of 5G and future technologies on infotainment connectivity is profound and far-reaching. With faster speeds, enhanced user experiences, and new business opportunities on the horizon, the future of infotainment is undoubtedly exciting and full of potential. As we continue to push the boundaries of what is possible, the infotainment industry will undoubtedly continue to evolve and adapt to the ever-changing landscape of connectivity and integration.

Challenges and Opportunities in Ensuring Seamless Infotainment Integration

As the infotainment landscape continues to evolve, the need for seamless integration between various devices, platforms, and services becomes increasingly crucial. This integration is essential for providing users with a cohesive and enjoyable experience. However, achieving this level of integration is not without its challenges. In this section, we will explore some of the key obstacles faced by the industry in ensuring seamless infotainment integration, as well as the opportunities that arise from overcoming these challenges.

One of the primary challenges in achieving seamless infotainment integration is the sheer diversity of devices and platforms available in the market. From smartphones and tablets to smart TVs and gaming consoles, users have a wide range of options when it comes to accessing infotainment content. Furthermore, each device runs on different operating systems, such as Android, iOS, and Windows, which adds another layer of complexity to the integration process.

To overcome this challenge, developers and service providers must adopt a cross-platform approach, ensuring their applications and content are compatible with multiple devices and operating systems. This may involve using universal programming languages like HTML5 or developing dedicated apps for each platform. By doing so, they can provide users with a consistent experience across all devices, ultimately enhancing user satisfaction and engagement.

As infotainment systems become more connected and integrated, concerns surrounding data privacy and security inevitably arise. Users are often required to share personal information, such as their location, preferences, and usage habits, to access personalized content and services. If not properly protected, this data can be vulnerable to breaches and misuse.

To address these concerns, industry players must prioritize data privacy and security in their integration efforts. This can be achieved by implementing robust encryption technologies, secure data storage solutions, and strict privacy policies. By protecting users' data, service providers can build trust and encourage greater adoption of their infotainment offerings.

A lack of standardization and interoperability between devices, platforms, and services is another challenge that hinders seamless infotainment integration. Without common standards and protocols, it becomes difficult for different systems to communicate and share information effectively.

Industry stakeholders must collaborate to develop and adopt universal standards and protocols that facilitate seamless integration. This may involve the creation of industry-wide consortiums or the adoption of existing standards, such as those developed by the World Wide Web Consortium (W3C). By fostering greater standardization and interoperability, the industry can pave the way for more seamless and user-friendly infotainment experiences.

Despite the challenges, there are significant opportunities for those who can successfully achieve seamless infotainment integration. By providing users with a cohesive and enjoyable experience across multiple devices and platforms, service providers can increase user engagement, drive customer loyalty, and generate higher revenues.

Moreover, seamless integration can open up new avenues for innovation and collaboration within the industry. As devices and services become more interconnected, there is potential for the development of new business models, partnerships, and revenue streams. For example, the integration of infotainment services with smart home systems could enable users to control their home environment through their entertainment devices, creating new opportunities for both industries.

In conclusion, the future of connectivity and integration in infotainment is full of challenges and opportunities. By addressing the obstacles related to device diversity, data privacy, and standardization, industry players can unlock the full potential of seamless infotainment experiences and capitalize on the opportunities that lie ahead.

The Future of Connectivity and Integration in Infotainment

As we have explored throughout this chapter, the world of infotainment is rapidly evolving, driven by advancements in technology and the ever-growing demand for seamless connectivity and integration. The future of infotainment is undoubtedly exciting, as we can expect even more immersive and personalized experiences that cater to our individual preferences and needs.

One of the key factors shaping the future of connectivity and integration in infotainment is the continued development of smart devices and the Internet of Things (IoT). As these technologies become more sophisticated and widespread, we can anticipate a more interconnected world where infotainment systems are seamlessly integrated with our daily lives. This will enable us to access information, entertainment, and services in real time, regardless of our location or device.

Another major driving force in the future of infotainment is the advent of 5G and other emerging technologies. With faster and more reliable connections, we can expect a significant improvement in the quality and variety of infotainment experiences. This will enhance existing services and pave the way for new and innovative applications that were previously unimaginable.

Despite the numerous benefits and opportunities that come with enhanced connectivity and integration, there are also challenges that need to be addressed. Ensuring data privacy and security will be of paramount importance as infotainment systems become more interconnected and store increasing amounts of personal information. Additionally, the industry will need to tackle issues related to cross-platform compatibility and standardization to ensure a smooth and consistent user experience across different devices and services.

In conclusion, the future of connectivity and integration in infotainment is bright, with endless possibilities for innovation and growth. As technology continues to advance rapidly, we can look forward to a world where infotainment systems are an integral part of our daily lives, providing us with a seamless and personalized experience that keeps us informed, entertained, and connected. By addressing the challenges and embracing the opportunities that lie ahead, the infotainment industry can continue to thrive and shape the way we interact with the world around us.

Chapter Summary

- The evolution of connectivity and integration in infotainment has transformed the way we access, consume, and interact with information and entertainment content, from radio broadcasts to the Internet of Things (IoT).
- Smart devices and IoT play a crucial role in infotainment integration, enabling a more personalized and immersive experience for users by connecting various devices and platforms.
- Cross-platform compatibility is essential for a seamless infotainment experience, requiring developers to ensure their applications and content are compatible with multiple devices and operating systems.

- The advent of 5G technology promises to revolutionize the infotainment industry by providing lightning-fast connectivity, enabling more immersive and interactive experiences, and fostering the rise of edge computing.
- Ensuring data privacy and security is a significant challenge in infotainment integration, requiring robust encryption technologies, secure data storage solutions, and strict privacy policies.
- Standardization and interoperability between devices, platforms, and services are necessary for seamless infotainment integration, requiring industry stakeholders to collaborate and adopt universal standards and protocols.
- Seamless infotainment integration offers significant opportunities for increased user engagement, customer loyalty, and revenue generation, as well as fostering innovation and collaboration within the industry.
- The future of connectivity and integration in infotainment is bright, with endless possibilities for innovation and growth, as long as the industry addresses challenges related to data privacy, cross-platform compatibility, and standardization.

6

INFOTAINMENT AND THE INTERNET OF THINGS (IOT)

A futuristic smart city, where various Internet of Things (IoT) devices seamlessly communicate with each other.

In today's fast-paced world, the demand for information and entertainment is at an all-time high. People are constantly seeking new ways to stay informed, engaged, and entertained, leading to the rise of infotainment – a blend of information and entertainment. Infotainment is a rapidly growing industry that encompasses various forms of media, including television, radio, print, and digital platforms. It aims to provide audiences with a captivating experience that combines the informative aspects of news and educational content with the entertaining elements of movies, music, and games.

As technology continues to advance, the Internet of Things (IoT) has emerged as a game-changer in the world of infotainment. IoT refers to the interconnection of everyday objects and devices through the internet, allowing them to communicate and exchange data with one another. This widespread connectivity has the potential to revolutionize the way we consume and interact with infotainment, making it more personalized, immersive, and accessible than ever before.

The integration of infotainment and IoT has opened up a world of possibilities, transforming the way we experience entertainment and access information. In this chapter, we will explore the evolution of infotainment systems in the IoT era, examining how these technologies are converging to create a seamless and interconnected experience for users. We will also discuss the integration of infotainment with smart home and smart city technologies, highlighting the potential for even greater personalization and customization of our entertainment experiences.

As with any technological advancement, the fusion of infotainment and IoT also raises concerns about security and privacy. We will delve into these issues, discussing the potential risks and challenges associated with IoT-based infotainment systems and the measures to address them.

Finally, we will look ahead to the future of infotainment in an interconnected world, considering the exciting possibilities and potential pitfalls that lie ahead as we continue to embrace the power of IoT. So, sit back, relax, and join us on this fascinating journey into the world of infotainment and the Internet of Things.

Evolution of Infotainment Systems in the IoT Era

The dawn of the Internet of Things (IoT) has brought about a significant transformation in the world of infotainment. As the IoT era unfolds, infotainment systems have evolved from being standalone entertainment devices to becoming an integral part of a connected ecosystem. This evolution has enhanced the user experience and opened up new avenues for innovation and growth in the infotainment industry.

In the early days, infotainment systems were primarily focused on providing entertainment and information through traditional mediums such as radio, television, and print. However, with the advent of the internet and the proliferation of smartphones, the infotainment landscape witnessed a paradigm shift. Users now had access to a plethora of content and services at their fingertips, leading to the emergence of new infotainment platforms such as social media, streaming services, and online news portals.

The IoT era has further revolutionized the infotainment industry by enabling seamless integration of infotainment systems with various connected devices and platforms. This has led to the development of advanced infotainment systems that offer users a more immersive and personalized experience. Some key milestones in the evolution of infotainment systems in the IoT era include:

Connected Cars: Modern vehicles are now equipped with sophisticated infotainment systems that offer a wide range of services such as navigation, real-time traffic updates, voice control, and smartphone integration. These systems leverage IoT technology to communicate with other connected devices, enabling features such as remote diagnostics, vehicle tracking, and over-the-air software updates.

Smart TVs: The advent of smart TVs has transformed the traditional television viewing experience by offering users access to a vast array of online content and services. Smart TVs now come with built-in

IoT capabilities, allowing users to control various smart home devices and access personalized content recommendations based on their viewing habits.

Voice Assistants: Voice-controlled digital assistants such as Amazon's Alexa, Google Assistant, and Apple's Siri have become integral to the infotainment ecosystem. These AI-powered assistants leverage IoT technology to interact with various connected devices, enabling users to access information, control smart home appliances, and even make purchases using simple voice commands.

Wearable Technology: Wearable devices such as smartwatches and fitness trackers have added a new dimension to the infotainment experience by offering users access to personalized content and services on the go. These devices leverage IoT technology to collect and analyze user data, providing insights into their health, fitness, and lifestyle habits.

As the IoT era continues to unfold, the infotainment landscape is poised for further growth and innovation. The seamless integration of infotainment systems with various connected devices and platforms will not only enhance the user experience but will also create new opportunities for content creators, service providers, and technology companies to thrive in this interconnected world.

Integration of Infotainment with Smart Home and Smart City Technologies

Rapid technological advancements have led to the seamless integration of infotainment systems with smart home and smart city technologies. This integration has not only enhanced the user experience but has also transformed the way we consume information and entertainment. In this section, we will delve into the various ways infotainment systems are being integrated into smart home and smart city technologies, and how this integration is shaping the future of infotainment.

Smart homes are equipped with advanced automation systems that allow homeowners to control various aspects of their homes, such as lighting, temperature, and security, using a single device or application.

Infotainment systems have become an integral part of smart homes, providing users with a centralized platform to access and control their entertainment options.

- **Centralized Control:** With the integration of infotainment systems in smart homes, users can now control their entertainment devices, such as televisions, gaming consoles, and audio systems, from a single interface. This centralized control simplifies the user experience and allows for seamless interaction between different devices.
- **Voice-activated Assistance:** The incorporation of voice-activated assistants like Amazon's Alexa, Google Assistant, and Apple's Siri into infotainment systems has made it easier for users to access and control their entertainment options. Users can now simply use voice commands to play music, watch movies, or even get the latest news updates.
- **Personalized Recommendations:** Infotainment systems in smart homes can now analyze user preferences and habits to provide personalized content recommendations. This ensures that users are always presented with relevant and engaging content, enhancing their overall entertainment experience.

Smart cities are urban areas that utilize advanced technology and data analysis to improve the quality of life for their residents. Infotainment systems play a crucial role in smart cities, providing residents with real-time information and entertainment options that cater to their needs and preferences.

- **Public Information Displays:** Infotainment systems integrated into smart city infrastructure can provide residents with real-time information on various aspects of city life, such as weather updates, traffic conditions, and public transportation schedules. These displays can be found in public spaces like parks, bus stops, and train stations, ensuring that residents have access to relevant information at all times.

- **Interactive Kiosks:** Smart cities are increasingly adopting interactive kiosks that provide residents with access to infotainment services. These kiosks can be used to access local news, events, and entertainment options, as well as provide directions and information about nearby attractions and amenities.
- **Connected Vehicles:** The integration of infotainment systems in connected vehicles has transformed the in-car entertainment experience. Drivers and passengers can now access real-time traffic updates, navigation services, and personalized entertainment options, making their journeys more enjoyable and efficient.

In conclusion, the integration of infotainment systems with smart home and smart city technologies has significantly enhanced the user experience, providing individuals with personalized and relevant content that caters to their needs and preferences. As technology continues to evolve, we can expect to see even more innovative ways in which infotainment systems are integrated into our daily lives, shaping the future of information and entertainment consumption.

Personalization and Customization of Infotainment Experiences through IoT

The Internet of Things (IoT) has revolutionized the way we interact with technology, allowing for the seamless integration of devices and systems in our daily lives. This interconnectedness has also profoundly impacted the world of infotainment, enabling users to enjoy highly personalized and customized experiences tailored to their preferences and needs. This section will explore how IoT has transformed the infotainment landscape, offering users a more immersive and engaging experience.

One of the most significant benefits of IoT in infotainment is the ability to collect and analyze vast amounts of user data. By gathering information on users' preferences, habits, and interests, IoT-enabled infotainment systems can create personalized content recommendations and experiences. For example, a smart TV connected to the IoT can analyze a user's viewing history and suggest movies or TV shows based on their preferences. Similarly, a connected car infotainment system can learn a driver's favorite routes, music preferences, and even preferred cabin temperature, ensuring a comfortable and enjoyable driving experience.

Another aspect of personalization and customization in IoT-based infotainment systems is the integration of voice assistants and natural language processing technologies. This allows users to interact with their devices using voice commands, making it easier to access content and control various aspects of the infotainment experience. For instance, a user can ask their smart speaker to play a specific song, provide news updates, or control smart home devices, all without lifting a finger. This level of convenience and personalization has become increasingly important as users seek more intuitive and user-friendly ways to interact with technology.

The IoT also enables infotainment systems to be more context-aware, adapting to the user's environment and situation. For example, a connected car infotainment system can detect when the vehicle is in heavy traffic and automatically switch to a more relaxing playlist or provide real-time traffic updates to help the driver navigate the congestion. Similarly, a smart home entertainment system can adjust the room's lighting and temperature based on the content consumed, creating a more immersive and enjoyable experience.

However, it is important to note that the personalization and customization of infotainment experiences through IoT come with certain challenges, particularly in terms of security and privacy. As these systems collect and store vast amounts of user data, there is an increased risk of data breaches and unauthorized access to sensitive information. Additionally, users may be concerned about how much their personal data is being used and shared by IoT-enabled infotainment systems. Addressing these concerns will be crucial in ensuring the continued growth and success of IoT-based infotainment solutions.

In conclusion, the Internet of Things has significantly transformed the world of infotainment, offering users highly personalized and customized experiences that cater to their unique preferences and needs. From data-driven content recommendations to context-aware adaptations, IoT-based infotainment systems are revolutionizing the way we consume and interact with content. As we continue to embrace the interconnected world of IoT, the future of infotainment promises to be more engaging, immersive, and tailored to the individual user than ever before.

Security and Privacy Concerns in IoT-based Infotainment Systems

As infotainment systems become increasingly interconnected with the Internet of Things (IoT), it is crucial to address the security and privacy concerns that arise from this integration. While IoT-based infotainment systems offer numerous benefits, such as personalized experiences and seamless integration with smart home and smart city technologies, they also present potential risks to users' personal information and overall system security. This section will explore the various security and privacy concerns associated with IoT-based infotainment systems and discuss potential solutions to mitigate these risks.

One of the primary concerns with IoT-based infotainment systems is the potential for unauthorized access to users' personal information. As these systems collect and store vast amounts of data, including users' preferences, location history, and even biometric data, it is essential to ensure this information remains secure and confidential. To address this concern, companies must implement robust data encryption and secure storage solutions, as well as adopt transparent data collection policies that inform users about the types of data being collected and how they will be used.

IoT-based infotainment systems often rely on multiple devices and networks to function seamlessly. This interconnectedness, however, also presents potential security vulnerabilities. Hackers may exploit weak points in the system to gain unauthorized access, potentially compromising the infotainment system and other connected devices and networks. To mitigate this risk, companies must prioritize device and network security by implementing strong authentication protocols, regularly updating software and firmware, and monitoring for potential threats.

As infotainment systems become more integrated with third-party applications and services, the potential for security breaches increases. Users must be cautious when granting permissions to these third-party integrations, as they may inadvertently expose their personal information or create vulnerabilities in the system. Companies should carefully vet third-party partners and ensure that they adhere to strict security and privacy standards.

A crucial aspect of addressing security and privacy concerns in IoT-based infotainment systems is educating users about potential risks and best practices for protecting their information. Companies should provide clear guidelines on how to create strong passwords, enable two-factor authentication, and regularly update software. Additionally, users should be encouraged to review and adjust privacy settings to ensure that they are comfortable with the level of data being collected and shared.

In conclusion, the integration of infotainment systems with IoT technologies presents both exciting opportunities and potential security and privacy risks. By prioritizing data protection, device and network security, and user education, companies can help ensure that users enjoy the benefits of IoT based infotainment systems while minimizing potential threats. As the IoT continues to evolve, industry leaders need to remain vigilant in addressing these concerns and fostering a secure and privacy-conscious environment for users.

The Future of Infotainment in an Interconnected World

As we stand at the precipice of a new era in technology, the future of infotainment in an interconnected world is both exciting and full of potential. The Internet of Things (IoT) has already begun to revolutionize the way we interact with our environment, and its integration with infotainment systems promises to create a seamless and personalized experience for users across various platforms.

The evolution of infotainment systems in the IoT era has seen a rapid transformation from standalone devices to interconnected ecosystems that communicate with each other and adapt to the user's preferences. This has allowed for a more immersive and engaging experience, as users can now access a wealth of information and entertainment options at their fingertips.

The integration of infotainment with smart home and smart city technologies has further expanded the possibilities for users, allowing them to control various aspects of their environment through a single interface. This simplifies the user experience and enables more efficient use of resources, ultimately contributing to a more sustainable future.

Personalization and customization of infotainment experiences through IoT have also become a key focus for developers as they strive to create systems that cater to each user's unique preferences and needs. This has led to the development of advanced algorithms and machine learning techniques that can analyze user behavior and adapt the infotainment system accordingly, ensuring a truly tailored experience.

However, the rapid expansion of IoT-based infotainment systems has also raised concerns about security and privacy. As more devices become interconnected, the potential for cyberattacks and data breaches increases, posing a significant risk to users. It is therefore crucial for developers and industry experts to prioritize the development of robust security measures and educate users about the importance of safeguarding their personal information.

In conclusion, the future of infotainment in an interconnected world is undoubtedly bright, with the potential to transform the way we consume information and entertainment. As IoT technology continues to evolve and integrate with infotainment systems, we can expect to see even more innovative and personalized experiences for users. However, we must remain vigilant about the potential risks associated with this rapid growth and ensure that security and privacy remain at the forefront of development. By striking the right balance between innovation and security, we can look forward to a truly interconnected world where infotainment systems enrich our lives and empower us to make informed decisions.

Chapter Summary

- Infotainment, a blend of information and entertainment, has evolved significantly with the advent of the Internet of Things (IoT), offering users a more personalized and immersive experience.
- IoT has enabled the integration of infotainment systems with various connected devices and platforms, leading to the development of advanced systems such as connected cars, smart TVs, voice assistants, and wearable technology.
- The integration of infotainment with smart home and smart city technologies has enhanced user experiences by providing centralized control, voice-activated assistance, and personalized recommendations.
- IoT allows for personalization and customization of infotainment experiences by collecting and analyzing user data, enabling context-aware adaptations and data-driven content recommendations.
- Security and privacy concerns arise with IoT-based infotainment systems due to the potential for unauthorized access to users' personal information and overall system security.
- To address security and privacy concerns, companies must implement robust data encryption, secure storage solutions, strong authentication protocols, and regular software updates.

- User awareness and education are crucial in addressing security and privacy concerns, with companies providing guidelines on creating strong passwords, enabling two-factor authentication, and adjusting privacy settings.
- The future of infotainment in an interconnected world is promising, with the potential to transform the way we consume information and entertainment. However, striking the right balance between innovation and security is essential to ensure a safe and privacy-conscious environment for users.

7

EMERGING TECHNOLOGIES IN INFOTAINMENT

A futuristic infotainment system seamlessly integrated into the dashboard of a sleek, self-driving car.

Infotainment has become a vital aspect of our daily lives. It refers to the fusion of informative and entertaining content delivered through various media platforms such as television, radio, print, and digital channels. The primary goal of infotainment is to educate and inform audiences while keeping them engaged and entertained. As technology continues to evolve rapidly, the infotainment industry is undergoing a significant transformation, with emerging technologies playing a crucial role in shaping its future.

In this chapter, we will delve into the world of emerging technologies in infotainment, exploring how innovations such as augmented and virtual reality, artificial intelligence, 5G, the Internet of Things (IoT), and voice assistants are revolutionizing the way we consume and interact with content. These technologies are enhancing the overall infotainment experience and opening up new avenues for content creators and providers to engage with their audiences in more immersive and personalized ways.

The rise of augmented and virtual reality has paved the way for more interactive and immersive infotainment experiences, allowing users to step into virtual worlds or overlay digital information onto their physical surroundings. Artificial intelligence and machine learning are revolutionizing content creation, enabling the development of more personalized and engaging content that caters to individual preferences and interests. The advent of 5G and IoT is set to transform the way we access and consume infotainment, providing seamless connectivity and enabling the integration of various smart devices for a more unified experience. Voice assistants and smart speakers are also gaining prominence in the infotainment landscape, offering users a more convenient and hands-free way to access content.

As we explore these emerging technologies in infotainment, it is essential to understand their transformative potential and the opportunities they present for both content creators and consumers. By embracing these innovations, the infotainment industry can continue to evolve and thrive in an increasingly digital and connected world, offering audiences more engaging, informative, and entertaining experiences than ever before.

The Rise of Augmented and Virtual Reality in Infotainment

In recent years, the infotainment landscape has been transformed by the rapid development and adoption of augmented reality (AR) and virtual reality (VR) technologies. These cutting-edge innovations have opened up new possibilities for content creators and consumers alike, allowing for immersive and interactive experiences that were once the stuff of science fiction. In this section, we will explore the rise of AR and VR in infotainment, examining their potential to revolutionize the way we consume and engage with content.

Augmented reality refers to the overlaying of digital information onto the user's real-world environment, typically through the use of a smartphone or wearable device. This technology has found a variety of applications in the infotainment sphere, from interactive museum exhibits to location-based games like Pokémon Go. By blending the digital and physical worlds, AR has the power to enrich our experiences and provide contextually relevant information in real time.

Virtual reality, on the other hand, transports users into a fully digital environment, often through the use of a head-mounted display (HMD) or VR headset. The entertainment industry has embraced this technology, with VR experiences ranging from immersive films and documentaries to interactive video games. In addition to providing a new level of immersion, VR has the potential to democratize access to experiences that were once limited by geographical or financial constraints, such as attending a live concert or exploring a far-off destination.

One of the most significant impacts of AR and VR on infotainment is the shift from passive consumption to active engagement. These technologies encourage users to become active participants in their experiences, whether it's by interacting with digital objects in an AR app or navigating a virtual environment in VR. This increased interactivity makes content more engaging and opens up new avenues for learning and exploration.

Moreover, AR and VR have the potential to create more personalized and tailored infotainment experiences. By leveraging user data and preferences, content creators can develop experiences that cater to individual interests and needs. For example, a virtual reality travel experience could be customized to highlight specific landmarks or activities based on the user's preferences, while an augmented reality museum tour could adapt its content to the user's level of expertise or interest in a particular subject.

In conclusion, the rise of augmented and virtual reality in infotainment marks a significant shift in the way we consume and engage with content. These emerging technologies have the power to transform passive experiences into interactive and immersive adventures while also offering the potential for more personalized and tailored content. As AR and VR continue to evolve and become more accessible, we can expect to see even more innovative and engaging infotainment experiences in the years to come.

Artificial Intelligence and Machine Learning: Revolutionizing Content Creation

In recent years, the infotainment industry has experienced a significant shift in the way content is created and consumed, thanks to the rapid advancements in artificial intelligence (AI) and machine learning (ML) technologies. These cutting-edge technologies have not only transformed the way we access and interact with information but also revolutionized the content creation process itself. In this section, we will delve into the impact of AI and ML on content creation and explore how these technologies are shaping the future of infotainment.

One of the most significant ways AI and ML have impacted content creation is by enabling the generation of personalized content. By analyzing user preferences, behaviors, and interests, AI-powered algorithms can create tailor-made content that caters to the specific needs and tastes of individual users. This level of personalization has led to a more engaging and immersive infotainment experience, as users are now presented with highly relevant and appealing content.

AI and ML have also paved the way for automated content production, which has the potential to revolutionize the way content is created and distributed. With the help of natural language processing (NLP) and natural language generation (NLG) technologies, AI-powered systems can now generate news articles, social media posts, and even creative writing pieces with minimal human intervention. This saves time and resources and allows content creators to focus on more strategic and creative aspects of their work.

The integration of AI and ML in content creation has also led to significant improvements in visual and audio content. For instance, AI-powered image and video editing tools can now automatically enhance the quality of visuals, remove unwanted elements, and even generate realistic images from scratch. Similarly, AI-driven audio processing tools can analyze and modify sound to create immersive audio experiences, such as 3D audio and adaptive soundtracks.

Another notable application of AI and ML in content creation is predictive content analytics. By analyzing historical data and user behavior patterns, AI-powered systems can predict which content is likely to resonate with users and generate higher engagement. This enables content creators to make data-driven decisions and optimize their content strategies for maximum impact.

While AI and ML have made significant strides in content creation, they are not meant to replace human creativity. Instead, these technologies can be seen as powerful tools that complement and enhance human creativity. By leveraging AI and ML, content creators can automate repetitive tasks, gain valuable insights, and focus on the more creative aspects of their work. This collaborative approach between AI and human creators has the potential to unlock new possibilities and push the boundaries of infotainment content.

In conclusion, the integration of artificial intelligence and machine learning in content creation has brought about a paradigm shift in the infotainment industry. By enabling personalized content generation, automating content production, enhancing visual and audio content, and providing predictive content analytics, AI and ML have revolutionized the way content is created and consumed. As these technologies continue to evolve, we can expect even more innovative and immersive infotainment experiences in the years to come.

The Impact of 5G and IoT on Infotainment Experiences

The world of infotainment is continuously evolving, and the introduction of 5G and the Internet of Things (IoT) is set to revolutionize the way we consume and interact with content. This section will explore how these emerging technologies transform infotainment experiences, enabling faster connectivity, seamless integration, and personalized content delivery.

The Power of 5G Connectivity

5G, the fifth generation of mobile networks, promises to deliver lightning-fast internet speeds, ultra-low latency, and increased network capacity. This technological leap will profoundly impact infotainment experiences, as it enables streaming of high-quality content without buffering or lag. Imagine watching a live sports event or concert in 4K resolution on your mobile device without any interruptions or delays – that's the power of 5G connectivity.

Moreover, 5G will enable the seamless integration of augmented and virtual reality experiences into infotainment platforms. With its low latency and high-speed data transfer capabilities, 5G will allow users to enjoy immersive AR and VR content without any performance issues, further blurring the lines between the digital and physical worlds.

The Internet of Things (IoT) and Infotainment

The Internet of Things (IoT) refers to the interconnection of everyday objects, such as appliances, vehicles, and wearable devices, through the internet. This technology has the potential to revolutionize the infotainment landscape by providing seamless integration and personalization of content across multiple devices.

For instance, IoT-enabled smart TVs can automatically recommend content based on your viewing habits and preferences, while smart speakers can curate playlists based on your mood or the time of day. In the automotive industry, connected cars can provide real-time traffic updates, navigation assistance, and personalized entertainment options, making your daily commute more enjoyable and efficient.

The Synergy of 5G and IoT in Infotainment

The combination of 5G and IoT technologies will create a powerful synergy that will redefine the infotainment experience. With 5G's high-speed connectivity and IoT's seamless integration capabilities, users can expect a more immersive, personalized, and interactive infotainment experience.

For example, imagine attending a live concert through a virtual reality headset, where you can choose your vantage point, interact with other attendees, and even influence the performance in real time. Or picture a smart home where your IoT devices work together to create a personalized infotainment experience, adjusting lighting, temperature, and audio settings to match the content you're consuming.

The impact of 5G and IoT on infotainment experiences cannot be overstated. These emerging technologies will enhance how we consume content and create new opportunities for content creators and service providers to deliver innovative and immersive experiences. As 5G and IoT continue to gain traction, we can expect a future where infotainment is more engaging, personalized, and integrated into our daily lives than ever before.

The Future of Voice Assistants and Smart Speakers in Infotainment

Rapid technological advancements have paved the way for voice assistants and smart speakers to become an integral part of our daily lives. These devices, powered by artificial intelligence, are revolutionizing the way we interact with technology and consume information. As the infotainment industry continues to evolve, voice assistants and smart speakers are poised to play a significant role in shaping the future of this domain.

Voice assistants, such as Amazon's Alexa, Apple's Siri, and Google Assistant, are becoming increasingly sophisticated, capable of understanding natural language and providing personalized responses. This has led to a surge in the popularity of smart speakers, which serve as a convenient and hands-free medium for accessing infotainment content. As these technologies continue to advance, we can expect them to become even more integrated into our infotainment experiences.

One of the most significant impacts of voice assistants and smart speakers on the infotainment industry is the shift toward audio-based content. Podcasts, audiobooks, and news briefings are gaining popularity as users seek convenient and engaging ways to consume information. This trend is expected to continue, with voice assistants and smart speakers becoming the primary medium for accessing and discovering audio content.

Moreover, voice assistants and smart speakers are transforming the way we interact with our devices, making them more natural and intuitive. As a result, content creators and providers need to adapt their strategies to cater to this new mode of interaction. This includes optimizing their content for voice search, creating voice-activated applications, and developing interactive experiences that leverage the unique capabilities of voice assistants.

Another exciting development in voice assistants and smart speakers is the potential for personalized and context-aware infotainment experiences. As these devices become more intelligent, they will be able to understand user preferences, habits, and contexts, allowing them to deliver tailored content that is both relevant and engaging. This level of personalization will enhance the user experience and create new opportunities for targeted advertising and content monetization.

In addition to personalization, the future of voice assistants and smart speakers in infotainment will also be shaped by the integration of other emerging technologies. For instance, the combination of voice assistants with augmented and virtual reality could lead to immersive and interactive experiences that redefine the way we consume infotainment content. Similarly, the convergence of voice assistants with IoT devices and 5G networks could enable seamless and interconnected infotainment experiences across various platforms and devices.

In conclusion, the future of voice assistants and smart speakers in infotainment is incredibly promising, with the potential to transform the way we access, discover, and engage with content. As these technologies continue to advance, we expect them to become even more integral to our infotainment experiences, driving innovation and creating new opportunities for content creators and consumers. Embracing these changes and adapting to the evolving landscape will be crucial for the infotainment industry to thrive in the era of voice assistants and smart speakers.

The Transformative Potential of Emerging Technologies in Infotainment

As we have explored throughout this chapter, the world of infotainment is undergoing a rapid and exciting transformation, thanks to the integration of emerging technologies. From augmented and virtual reality to artificial intelligence and machine learning, these advancements are revolutionizing the way we create, consume, and interact with content. As we look to the future, it is clear that these technologies hold immense potential to reshape the infotainment landscape, offering new opportunities for innovation and growth.

One of the most significant impacts of these emerging technologies is the democratization of content creation. With tools like AI and machine learning, creators can now develop personalized, immersive experiences that cater to individual preferences and needs. This enhances the overall quality of infotainment and fosters a more inclusive and diverse ecosystem where a wider range of voices and perspectives can be heard.

Furthermore, the advent of 5G and IoT technologies can revolutionize how we access and engage with infotainment. As these technologies become more widespread, we can expect to see a seamless integration of infotainment experiences into our daily lives, with content being delivered faster and more efficiently than ever before. This will enable us to stay connected and informed, no matter where we are or what we are doing.

Voice assistants and smart speakers are another area of growth in the infotainment sector, as they offer a convenient and accessible way for users to access content. As these devices become more sophisticated and capable of understanding natural language, we can expect to see a shift towards voice-based interactions, making infotainment experiences more intuitive and user-friendly.

In conclusion, the transformative potential of emerging technologies in infotainment is undeniable. As we continue to push the boundaries of what is possible, we can look forward to a future where infotainment is more immersive, personalized, and accessible than ever before. By embracing these innovations, we can unlock new opportunities for growth and development, ensuring that infotainment remains a vital and dynamic part of our lives for years to come.

Chapter Summary

- Infotainment, the fusion of informative and entertaining content, is undergoing a significant transformation due to emerging technologies such as augmented and virtual reality, artificial intelligence, 5G, IoT, and voice assistants.
- Augmented and virtual reality are revolutionizing infotainment by providing more interactive and immersive experiences, allowing users to actively engage with content and enabling personalized experiences based on individual preferences.
- Artificial intelligence and machine learning are transforming content creation by enabling personalized content generation, automating content production,

enhancing visual and audio content, providing predictive content analytics, and fostering collaboration between AI and human creators.

- The advent of 5G and IoT technologies is set to transform infotainment experiences by providing faster connectivity, seamless integration, and personalized content delivery across multiple devices.
- Voice assistants and smart speakers are becoming increasingly integral to infotainment experiences, driving a shift towards audio-based content and transforming the way users interact with devices.
- Personalization and context-aware infotainment experiences will be a key focus in the future, with voice assistants and smart speakers becoming more intelligent and capable of understanding user preferences, habits, and contexts.
- The integration of emerging technologies such as augmented and virtual reality, IoT devices, and 5G networks with voice assistants and smart speakers will enable seamless and interconnected infotainment experiences across various platforms and devices.
- The transformative potential of emerging technologies in infotainment is immense, offering new opportunities for innovation, growth, and the democratization of content creation, ultimately leading to more immersive, personalized, and accessible infotainment experiences.

8

CYBERSECURITY AND PRIVACY IN INFOTAINMENT SYSTEMS

A futuristic car infotainment system with advanced cybersecurity features.

In today's hyper-connected world, infotainment systems have become an integral part of our daily lives. These systems, which combine information and entertainment, are found in various forms, such as in-car entertainment systems, smart TVs, and streaming platforms. As the popularity of infotainment systems continues to grow, so does the need for robust cybersecurity and privacy measures to protect users and their data.

Cybersecurity and privacy in infotainment systems are crucial aspects that ensure the safety and security of users' personal information, as well as the proper functioning of the systems themselves. With the increasing amount of sensitive data being stored and transmitted through these systems, the risk of cyberattacks and data breaches has never been higher. This chapter aims to provide an overview of the importance of cybersecurity and privacy in infotainment systems, the common threats and vulnerabilities faced by these systems, and the best practices for ensuring their security and privacy.

As infotainment systems become more sophisticated and interconnected, they also become more vulnerable to cyber threats. Hackers and cybercriminals are constantly looking for ways to exploit vulnerabilities in these systems, which can lead to unauthorized access, data theft, and even physical harm to users. For instance, a hacker could potentially gain control of a vehicle's infotainment system and manipulate its functions, putting the driver and passengers at risk.

Moreover, privacy concerns have become increasingly relevant in the context of infotainment systems. Users often provide personal information, such as their location, preferences, and payment details, to access various features and services offered by these systems. If not properly protected, this data can be misused by malicious actors for identity theft, financial fraud, or other nefarious purposes.

In light of these risks, manufacturers, developers, and users of infotainment systems need to prioritize cybersecurity and privacy. By implementing strong security measures and adhering to industry standards and regulations, we can ensure that infotainment systems remain a safe and enjoyable part of our lives. In the following sections, we will delve deeper into the various aspects of cybersecurity and privacy in infotainment systems, providing insights and recommendations for a secure and private user experience.

The Growing Importance of Cybersecurity in Infotainment Systems

As infotainment systems continue to evolve and become more integrated into our daily lives, the importance of cybersecurity in these systems cannot be overstated. With the rapid advancement of technology, infotainment systems have transformed from simple radio and CD players to sophisticated, connected devices that offer a wide range of services such as navigation, communication, and entertainment. These systems are now essential to modern vehicles, smart homes, and personal devices, making them an attractive target for cybercriminals.

There are several factors contributing to the growing importance of cybersecurity in infotainment systems:

- **Increased Connectivity:** The integration of infotainment systems with the internet and other connected devices has opened up new avenues for potential cyberattacks. Hackers can exploit vulnerabilities in these systems to gain unauthorized access to sensitive information, manipulate system functionality, or even take control of the device.
- **Data Collection and Storage:** Modern infotainment systems collect and store a vast
- amount of personal data, such as user preferences, location history, and communication logs. This data can be highly valuable to cybercriminals, who may use it for identity theft, targeted advertising, or other malicious purposes.
- **Complexity of Systems:** The complexity of infotainment systems has increased exponentially in recent years with the addition of new features, applications, and interfaces. This complexity makes it more challenging to identify and

address potential security vulnerabilities, leaving the systems more susceptible to cyberattacks.

- **User Trust and Expectations:** As users become more reliant on infotainment systems for their daily needs, they expect these systems to be secure and protect their privacy. Any security or privacy breach can lead to a loss of trust in the system and its manufacturer, resulting in reputational damage and potential legal consequences.
- **Interconnectivity with Critical Systems:** In the case of automotive infotainment systems, the increasing integration with other vehicle systems, such as engine control and safety features, raises the stakes for cybersecurity. A successful cyberattack on an infotainment system could potentially compromise the safety and functionality of the entire vehicle.

In light of these factors, it is crucial for manufacturers, developers, and users of infotainment systems to prioritize cybersecurity and privacy. By understanding the risks and implementing robust security measures, we can ensure that infotainment systems continue to provide the convenience and entertainment we have come to expect while safeguarding our personal information and safety.

Common Cybersecurity Threats and Vulnerabilities in Infotainment Systems

As infotainment systems become more advanced and interconnected, they also become more susceptible to various cybersecurity threats and vulnerabilities. In this section, we will explore some of the most common threats and vulnerabilities that infotainment systems face today, and how they can potentially impact users' privacy and security.

One of the primary concerns in infotainment systems is unauthorized access to sensitive data. This can occur through various means, such as hacking into the system, exploiting software vulnerabilities, or intercepting data transmissions. Data breaches can lead to the exposure of personal information, such as user preferences, location data, and even payment details, which can then be used for identity theft, fraud, or other malicious activities.

Malware and ransomware attacks are becoming increasingly common in infotainment systems. Cybercriminals can use these attacks to gain control of a user's infotainment system, lock them out, or even demand a ransom to restore access. In some cases, malware can also be used to track a user's location, monitor their activities, or steal sensitive information.

As infotainment systems become more complex, they also become more prone to software vulnerabilities and exploits. Hackers can exploit these vulnerabilities to gain unauthorized access to the system, manipulate its functions, or even take control of the vehicle itself. In some cases, these vulnerabilities can be discovered and exploited by cybercriminals before manufacturers have the chance to issue a patch or update, leaving users at risk.

Many infotainment systems rely on wireless communication technologies, such as Wi-Fi, Bluetooth, and cellular networks, to provide connectivity and functionality. However, these wireless connections can also be vulnerable to eavesdropping, interception, or tampering. This can lead to the exposure of sensitive data, unauthorized access to the infotainment system, or even the spread of malware between connected devices.

As infotainment systems become more open and customizable, users often have the option to install third-party applications and services. While this can provide additional functionality and convenience, it can also introduce new security risks. Some third-party applications may contain vulnerabilities or malicious code, which can compromise the security of the infotainment system and potentially lead to data breaches or other cyberattacks.

In conclusion, the growing complexity and connectivity of infotainment systems have led to an increased risk of cybersecurity threats and vulnerabilities. Therefore, manufacturers, developers, and users need to be aware of these risks and take appropriate measures to protect their systems and personal information. The following section will discuss best practices for ensuring privacy and security in infotainment systems.

Best Practices for Ensuring Privacy and Security in Infotainment Systems

As infotainment systems continue to evolve and become more integrated into our daily lives, it is crucial to prioritize privacy and security. This section will outline some best practices that can help ensure the safety and privacy of users while enjoying the benefits of infotainment systems.

Regular Software Updates: One of the most effective ways to protect infotainment systems from cyber threats is to keep the software up-to-date. Manufacturers should release regular updates to fix any security vulnerabilities and improve the overall performance of the system. Users should also be proactive in installing these updates as soon as they become available.

Strong Authentication Measures: Implementing strong authentication measures, such as multi-factor authentication or biometric identification, can significantly reduce the risk of unauthorized access to infotainment systems. This is particularly important for systems that store sensitive personal information or provide access to financial transactions.

Data Encryption: Encrypting data stored on infotainment systems can help protect user privacy and prevent unauthorized access to sensitive information. Manufacturers should use robust encryption algorithms and ensure that encryption keys are securely managed.

Secure Communication Channels: Infotainment systems often communicate with external devices and networks, making it essential to secure these communication channels. This can be achieved by using secure protocols, such as HTTPS and SSL/TLS, and regularly updating security certificates.

Privacy by Design: Manufacturers should adopt a privacy-by-design approach when developing infotainment systems. This means considering privacy and security implications from the initial design stages and incorporating appropriate safeguards throughout the development process.

User Education and Awareness: Users play a critical role in maintaining the privacy and security of infotainment systems. Manufacturers and service providers should provide clear guidance on best practices for using these systems safely and securely. This may include instructions on setting strong passwords, enabling security features, and recognizing potential phishing attempts.

Third-Party Security Audits: Regular third-party security audits can help identify potential vulnerabilities in infotainment systems and ensure that manufacturers adhere to industry best practices. These audits should be conducted by reputable cybersecurity firms with experience in the infotainment industry.

Incident Response Planning: Manufacturers should have a well-defined incident response plan in place to address any potential security breaches or privacy violations. This plan should outline the steps to be taken in the event of an incident, including notifying affected users, mitigating the impact, and preventing future occurrences.

By following these best practices, manufacturers and users can work together to create a safer and more secure infotainment ecosystem. As technology continues to advance, it is essential to remain vigilant and proactive in addressing potential cybersecurity and privacy risks. By doing so, we can ensure that infotainment systems continue to provide valuable and enjoyable experiences while protecting the privacy and security of users.

The Role of Legislation and Industry Standards in Protecting Infotainment Systems

As infotainment systems continue to evolve and become more sophisticated, the need for robust cybersecurity measures and privacy protection becomes increasingly critical. In this section, we will explore the role of legislation and industry standards in safeguarding infotainment systems and ensuring users can enjoy these technologies without compromising their security and privacy.

Legislation is crucial in setting the legal framework for cybersecurity and privacy protection in infotainment systems. Governments around the world have recognized the importance of securing these systems and have enacted laws and regulations to ensure that manufacturers and service providers adhere to certain security and privacy standards. For example, the European Union's General Data Protection Regulation (GDPR) has significantly impacted how companies handle user data, including data collected and processed by infotainment systems. In the United States, the Federal Trade Commission (FTC) enforces privacy and data security requirements for connected devices, including infotainment systems.

Industry standards also play a vital role in shaping the security and privacy landscape for infotainment systems. Organizations like the International Organization for Standardization (ISO) and the Society of Automotive Engineers (SAE) have developed guidelines and best practices for securing automotive systems, including infotainment systems. For instance, the ISO/SAE 21434 standard provides a comprehensive framework for addressing cybersecurity risks in the automotive industry, covering aspects such as risk management, vulnerability assessment, and incident response.

These industry standards serve as a benchmark for manufacturers and service providers, ensuring they follow best practices in securing their infotainment systems. By adhering to these standards, companies can demonstrate their commitment to cybersecurity and privacy, thereby building trust with consumers and regulatory authorities.

In addition to legislation and industry standards, collaboration between various stakeholders is essential in protecting infotainment systems. This includes cooperation between manufacturers, suppliers, software developers, and cybersecurity experts to share information about potential threats and vulnerabilities, as well as to develop and implement effective security measures. Public-private partnerships, such as the Automotive Information Sharing and Analysis Center (Auto-ISAC), facilitate this collaboration by providing a platform for sharing cybersecurity information and best practices among industry stakeholders.

In conclusion, legislation and industry standards play a pivotal role in ensuring the security and privacy of infotainment systems. By establishing a legal framework and setting benchmarks for best practices, these measures help to protect users from potential cybersecurity threats and privacy breaches. As infotainment systems continue to advance and become more interconnected, it is essential for all stakeholders to work together to stay ahead of emerging risks and safeguard the future of these technologies.

The Future of Cybersecurity and Privacy in Infotainment Systems

As we reach the end of this chapter, it is important to reflect on the significance of cybersecurity and privacy in the rapidly evolving world of infotainment systems. The integration of technology into our daily lives has brought about a plethora of opportunities for entertainment, communication, and convenience. However, with these advancements come the inevitable challenges of ensuring the security and privacy of users' data and personal information.

The future of cybersecurity and privacy in infotainment systems will be shaped by the continuous development of new technologies and the growing awareness of the potential risks associated with their use. As the industry moves forward, it is crucial for manufacturers, developers, and users alike to prioritize the protection of sensitive information and the prevention of cyber threats.

One of the key factors in achieving this goal is the implementation of robust security measures and best practices throughout the design and development process. By incorporating security features such as encryption, authentication, and regular software updates, manufacturers can significantly reduce the likelihood of cyberattacks and data breaches.

Moreover, the role of legislation and industry standards will continue to be critical in shaping the future of cybersecurity and privacy in infotainment systems. Governments and regulatory bodies must work together to establish comprehensive guidelines and requirements that promote the development of secure and privacy-respecting technologies. This collaborative effort will not only help to protect users' data but also foster trust and confidence in the infotainment industry as a whole.

In addition to these measures, the future of cybersecurity and privacy in infotainment systems will also be influenced by users' growing awareness and involvement. As consumers become more educated about the potential risks associated with their devices, they will be better equipped to make informed decisions about the products they use and the security measures they implement. This increased vigilance will ultimately contribute to the overall security and privacy of infotainment systems.

In conclusion, the future of cybersecurity and privacy in infotainment systems is a dynamic and multifaceted landscape that will be shaped by the ongoing collaboration between manufacturers, developers, governments, and users. By prioritizing security and privacy throughout the design, development, and implementation process, the industry can continue to innovate and evolve while ensuring the protection of users' data and personal information. As we move forward, it is essential to remain vigilant and proactive in addressing the ever-changing challenges and opportunities presented by the world of infotainment technology.

Chapter Summary

- The growing importance of cybersecurity in infotainment systems is due to increased connectivity, data collection and storage, complexity of systems, user trust and expectations, and interconnectivity with critical systems.
- Common cybersecurity threats and vulnerabilities in infotainment systems include unauthorized access and data breaches, malware and ransomware attacks, software vulnerabilities and exploits, insecure wireless communications, and third-party applications and services.

- Best practices for ensuring privacy and security in infotainment systems include regular software updates, strong authentication measures, data encryption, secure communication channels, privacy by design, user education and awareness, third-party security audits, and incident response planning.
- Legislation and industry standards play a crucial role in setting the legal framework for cybersecurity and privacy protection in infotainment systems, with examples such as the European Union's General Data Protection Regulation (GDPR) and the ISO/SAE 21434 standard.
- Collaboration between various stakeholders, including manufacturers, suppliers, software developers, and cybersecurity experts, is essential in protecting infotainment systems and sharing information about potential threats and vulnerabilities.
- Public-private partnerships, such as the Automotive Information Sharing and Analysis Center (Auto-ISAC), facilitate collaboration by providing a platform for sharing cybersecurity information and best practices among industry stakeholders.
- The future of cybersecurity and privacy in infotainment systems will be shaped by the continuous development of new technologies, growing awareness of potential risks, and the implementation of robust security measures and best practices throughout the design and development process.
- The ongoing collaboration between manufacturers, developers, governments, and users is essential for prioritizing security and privacy in infotainment systems, fostering trust and confidence in the industry, and ensuring the protection of users' data and personal information.

9

INFOTAINMENT AND ITS IMPACT ON SOCIETY

A futuristic smart city, where various Internet of Things (IoT) devices seamlessly communicate with each other.

In today's fast-paced and interconnected world, the media landscape has undergone a significant transformation. The traditional boundaries between news and entertainment have become increasingly blurred, giving rise to a new phenomenon known as infotainment. This term refers to the fusion of informative content with elements of entertainment, often presented in a sensationalized or attention-grabbing manner. As the demand for engaging and easily digestible content continues to grow, infotainment has become a prominent fixture in modern media, shaping the way we consume information and influencing our perceptions of the world around us.

The rise of infotainment can be attributed to several factors, including the proliferation of media outlets, the advent of the 24-hour news cycle, and the increasing influence of social media and technology. With an ever-growing number of channels and platforms vying for audience attention, media organizations have been compelled to adopt new strategies to attract and retain viewers. Infotainment has become a popular solution, offering a more engaging and accessible alternative to traditional news formats.

This shift towards infotainment has been further fueled by the rapid advancement of technology and the widespread adoption of social media. As smartphones and other devices have become increasingly ubiquitous, consumers now have instant access to a vast array of information and entertainment options at their fingertips. This has led to a significant change in audience expectations, with many people now seeking out content that is not only informative but also entertaining and shareable. In response, media organizations have increasingly turned to infotainment to cater to these evolving preferences.

While the rise of infotainment has undoubtedly transformed the way we consume information, it has also raised important questions about the impact of this phenomenon on society. As the lines between news and entertainment continue to blur, concerns have been raised about the potential consequences for public opinion, political discourse, and the ethical responsibilities of media providers. In this chapter, we will explore these issues in greater depth, examining the various ways in which infotainment has shaped our world and considering the implications for the future of media and society.

The Blurring Lines Between News and Entertainment

In recent years, the distinction between news and entertainment has become increasingly blurred, giving rise to the phenomenon of infotainment. This fusion of information and entertainment has led to a shift in the way people consume and perceive news, as well as the manner in which media outlets present it. In this section, we will delve into the factors contributing to this blurring of lines and the implications it has on society.

One of the primary reasons for the blurring of lines between news and entertainment is the fierce competition among media outlets for viewership and advertising revenue. In an attempt to attract a larger audience, news organizations have increasingly adopted sensationalist tactics, focusing on stories that are more entertaining and attention-grabbing rather than those of significant public interest. This has led to a shift in news content, with a greater emphasis on celebrity gossip, scandals, and human-interest stories, often at the expense of in-depth reporting on important issues.

Another contributing factor is the rise of the 24-hour news cycle, which has created a constant demand for fresh content. In order to fill this void, news organizations often resort to presenting stories in a more entertaining and engaging manner, with flashy visuals, dramatic storytelling, and even humorous commentary. This approach not only blurs the lines between news and entertainment but also has the potential to trivialize important issues and undermine the credibility of news organizations.

The influence of infotainment is also evident in the increasing prominence of opinion-driven news programming, where the focus is on providing commentary and analysis rather than objective reporting. This format often features charismatic personalities who express strong opinions and engage in heated debates, making for compelling television. However, this approach can contribute to the polarization of public opinion, as viewers are more likely to be exposed to biased perspectives that align with their own beliefs.

Moreover, the advent of social media and technology has played a significant role in the blurring of lines between news and entertainment. Platforms like Facebook and Twitter have become primary sources of news for many people, who often encounter news stories mixed in with entertaining content, such as memes, videos, and personal updates from friends. This blending of news and entertainment on social media platforms can make it difficult for users to distinguish between factual reporting and opinion, leading to the spread of misinformation and the erosion of trust in traditional news sources.

In conclusion, the blurring of lines between news and entertainment has significant implications for society, as it affects the way people consume and perceive news, as well as the manner in which media outlets present it. As infotainment continues to gain prominence in modern media, it is crucial for both consumers and providers of news to be aware of the potential consequences and to strive for a balance between engaging content and responsible journalism.

The Influence of Infotainment on Public Opinion and Political Discourse

The rise of infotainment has significantly impacted public opinion and political discourse. As the lines between news and entertainment continue to blur, the way people consume information and form opinions has evolved. This section will explore the ways in which infotainment influences public opinion and political discourse, and the potential consequences of this shift.

One of the most apparent effects of infotainment is the simplification of complex issues. Infotainment often presents information in an easily digestible and entertaining manner, which can lead to oversimplification. This can result in the public forming opinions based on incomplete or inaccurate information. For example, a news segment that focuses on the personal life of a political figure rather than their policies may lead viewers to form opinions based on superficial factors rather than the substance of their political beliefs.

Infotainment has also contributed to the rise of sensationalism in news reporting. In an effort to attract viewers and generate revenue, infotainment providers often prioritize stories that are shocking or emotionally charged. This can lead to an overemphasis on scandal and controversy, which may distort public opinion and overshadow more important issues. The focus on sensationalism can also contribute to a more polarized political climate, as people are exposed to extreme viewpoints and inflammatory rhetoric.

The influence of infotainment on political discourse is further exacerbated by the role of social media and technology. As people increasingly rely on social media platforms for news and information, they are exposed to a constant stream of infotainment content. This can create echo chambers, where individuals are only exposed to information that aligns with their beliefs and opinions. This phenomenon can lead to increased political polarization and a breakdown in civil discourse as people become less willing to engage with opposing viewpoints.

Moreover, the prevalence of infotainment on social media platforms can contribute to the spread of misinformation and fake news. As infotainment providers prioritize engagement and virality, they may be more likely to share misleading or factually incorrect content. This can have serious consequences for public opinion and political discourse, as people base their opinions on false information.

In conclusion, the influence of infotainment on public opinion and political discourse is multifaceted and complex. The simplification of complex issues, the rise of sensationalism, and the role of social media and technology have all contributed to a shift in the way people consume information and form opinions. As infotainment continues to grow in popularity, it is crucial for both consumers and providers to be aware of the potential consequences and to engage in responsible and ethical media consumption.

The Role of Social Media and Technology in the Spread of Infotainment

The rapid advancement of technology and the widespread use of social media platforms have played a significant role in the proliferation of infotainment. In this digital age, people are constantly connected to the internet, seeking instant gratification and easy access to information. As a result, traditional media outlets have had to adapt to the changing landscape, often incorporating infotainment elements into their content to maintain audience engagement.

Social media platforms, such as Facebook, Twitter, and Instagram, have become powerful tools for the dissemination of infotainment. These platforms allow users to share and consume content at an unprecedented rate, often blurring the lines between news and entertainment. The algorithms used by these platforms prioritize content that generates high levels of engagement, which often means that sensationalized, emotionally charged, or controversial infotainment stories are more likely to appear in users' feeds.

This phenomenon has led to the rise of "clickbait" headlines and sensationalized content, designed to capture users' attention and encourage them to share the content with their networks. In turn, this has created an environment where infotainment thrives, as it is often more engaging and shareable than traditional news content.

Moreover, the ease with which users can create and share content on social media has given rise to a new generation of infotainment providers. Influencers, bloggers, and even ordinary users can now create content that reaches millions of people, often without the need for traditional journalistic credentials or editorial oversight. This democratization of content creation has led to an influx of infotainment, as users are drawn to the entertaining and easily digestible nature of such content.

In addition to social media, technology has also facilitated the spread of infotainment by developing new content formats and platforms. Podcasts, streaming services, and mobile applications have all contributed to the growth of infotainment by providing users with on-demand access to content that combines news and entertainment. These platforms have also given rise to a new breed of infotainment providers who are able to reach large audiences without the constraints of traditional media outlets.

While the role of social media and technology in the spread of infotainment has undoubtedly increased its reach and influence, it has also raised concerns about the quality and accuracy of the information being consumed. The ease with which false or misleading information can be shared and spread on social media platforms has led to the rise of "fake news" and misinformation, which can have serious consequences for public opinion and political discourse.

In conclusion, the role of social media and technology in the spread of infotainment has been both a driving force behind its growth and a source of concern for its potential impact on society. As infotainment continues to evolve and adapt to the changing media landscape, both content creators and consumers need to be aware of the ethical implications and responsibilities associated with this form of media.

The Ethical Implications and Responsibilities of Infotainment Providers

As infotainment continues to gain prominence in modern media, it is crucial to examine the ethical implications and responsibilities that come with providing this type of content. Infotainment providers, including television networks, online platforms, and social media influencers, must navigate a complex landscape of ethical considerations to ensure that they are not compromising the integrity of information or contributing to the spread of misinformation.

One of the primary ethical concerns surrounding infotainment is the potential for sensationalism and the prioritization of entertainment value over factual accuracy. In the pursuit of higher ratings and increased engagement, infotainment providers may be tempted to exaggerate or distort the truth to create more compelling content. This can lead to the dissemination of misleading or false information, which can have serious consequences for public understanding and decision-making. To uphold ethical standards, infotainment providers must prioritize accuracy and transparency, ensuring that their content is grounded in verifiable facts and that any potential biases or inaccuracies are clearly acknowledged.

Another ethical consideration for infotainment providers is the potential for the trivialization of important issues. By presenting complex topics in a simplified and entertaining manner, infotainment can sometimes reduce the seriousness of these issues and contribute to a lack of public engagement with critical social, political, and environmental challenges. To address this concern, infotainment providers should strive to balance entertainment value with educational content, ensuring that their audience is informed and engaged.

The role of advertising and commercial interests in infotainment also raises ethical questions. Infotainment providers must be cautious not to allow the pursuit of profit to compromise the quality and integrity of their content. This includes avoiding sensationalism, clickbait tactics, and the promotion of products or services that may be harmful or misleading. To maintain ethical standards, infotainment providers should be transparent about their financial relationships and ensure that any sponsored content is clearly labeled as such.

In addition to these ethical considerations, infotainment providers have a responsibility to promote media literacy among their audience. As the lines between news and entertainment continue to blur, consumers must be equipped with the skills to critically evaluate the information they encounter. Infotainment providers can contribute to this effort by encouraging critical thinking, fostering open dialogue, and providing resources for further learning.

In conclusion, the ethical implications and responsibilities of infotainment providers are multifaceted and complex. By prioritizing accuracy, transparency, and media literacy, these providers can help to ensure that infotainment serves as a valuable and responsible source of information and entertainment for society. As the landscape of media continues to evolve, it is essential that infotainment providers remain vigilant in upholding these ethical standards and fostering a media environment that is both engaging and informative.

The Future of Infotainment and its Effects on Society

As we have explored throughout this chapter, infotainment has become an increasingly prevalent and influential force in modern media. The blending of news and entertainment has led to a shift in the way people consume information, with many prioritizing entertainment value over factual accuracy. This phenomenon has significantly impacted public opinion, political discourse, and the ethical responsibilities of media providers. In this conclusion, we will consider the future of infotainment and its potential effects on society.

The rise of infotainment can be attributed to various factors, including the rapid advancement of technology and the widespread use of social media platforms. As these trends continue to evolve, infotainment will likely become even more deeply ingrained in our daily lives. The proliferation of smartphones, tablets, and other digital devices has made it easier than ever for people to access and share content, and this trend shows no signs of slowing down. As a result, we can expect infotainment to become even more pervasive and influential in the coming years.

One potential outcome of this increased reliance on infotainment is a further blurring of the lines between news and entertainment. As audiences continue to prioritize engaging content over objective reporting, media providers may feel compelled to cater to these preferences by producing more sensationalized and opinion-driven content. This could lead to a decline in the quality and accuracy of news reporting, with potentially serious consequences for public opinion and political discourse.

On the other hand, the growing influence of infotainment could also lead to positive developments in the media landscape. For example, the increased competition among content providers may encourage innovation and creativity in the way news is presented and disseminated. This could result in new formats and platforms that offer audiences a more engaging and informative experience, ultimately contributing to a more informed and engaged public.

In order to mitigate the potential negative effects of infotainment on society, it is crucial for media providers to recognize and uphold their ethical responsibilities. This includes ensuring that news content is accurate, balanced, and transparent, even when it is presented in an entertaining format. Additionally, media providers should strive to educate their audiences about the importance of critical thinking and media literacy, empowering them to discern between fact and fiction and make informed decisions based on reliable information.

In conclusion, the future of infotainment and its effects on society will largely depend on the choices made by both media providers and consumers. By prioritizing ethical reporting and promoting media literacy, we can harness the power of infotainment to create a more informed and engaged public while minimizing its potential negative consequences. As the lines between news and entertainment continue to blur, all stakeholders need to remain vigilant and proactive in navigating this complex and ever-evolving media landscape.

Chapter Summary

- Infotainment, a fusion of information and entertainment, has become a prominent fixture in modern media due to factors such as the proliferation of media outlets, the 24-hour news cycle, and the influence of social media and technology.
- The blurring lines between news and entertainment have led to a shift in news content, with a greater emphasis on sensationalism, celebrity gossip, and human-interest stories, often at the expense of in-depth reporting on important issues.
- Infotainment significantly impacts public opinion and political discourse, as it can lead to the simplification of complex issues, the rise of sensationalism, and the creation of echo chambers on social media platforms.
- Social media and technology have played a significant role in the spread of infotainment, with platforms like Facebook

and Twitter becoming primary news sources for many people, often blurring the lines between factual reporting and opinion.

- The ethical implications and responsibilities of infotainment providers include prioritizing accuracy and transparency, avoiding sensationalism and clickbait tactics, and promoting media literacy among their audience.
- The future of infotainment is likely to see a further blurring of the lines between news and entertainment, with potential consequences for public opinion and political discourse.
- To mitigate the potential negative effects of infotainment on society, media providers must recognize and uphold their ethical responsibilities, ensuring that news content is accurate, balanced, and transparent.
- The future of infotainment and its effects on society will largely depend on the choices made by both media providers and consumers, with a focus on ethical reporting and promoting media literacy to create a more informed and engaged public.

THE INFOTAINMENT REVOLUTION

As we reach the end of our journey exploring the fascinating world of infotainment, we must reflect on the insights and knowledge we have gained throughout this book. The infotainment revolution has undeniably transformed how we consume information and entertainment, merging these two seemingly distinct realms into a unified experience. In this final chapter, we will unveil the infotainment epilogue, providing a comprehensive summary of the major themes and findings discussed in the preceding chapters and offering a glimpse into the future of this dynamic industry.

The infotainment landscape has evolved rapidly over the past few decades, driven by advancements in technology and the ever-changing preferences of consumers. From the early days of radio and television broadcasting to the rise of the internet and social media platforms, infotainment has consistently adapted to the needs and desires of its audience. This adaptability has allowed infotainment to permeate nearly every aspect of our lives, from the way we learn about current events to the manner in which we interact with our favorite celebrities and brands.

Throughout this book, we have delved into the various facets of the infotainment industry, examining its history, development, and impact on society. We have also explored the diverse range of infotainment formats and platforms, including news, television, film, music, gaming, and social media. By understanding the intricacies of these different mediums, we have gained a deeper appreciation for the complex and multifaceted nature of infotainment.

In this concluding chapter, we will revisit the major themes and findings that have emerged from our exploration of the infotainment landscape, highlighting the implications and significance of these insights for modern society. We will also address the limitations and critiques of the infotainment industry, acknowledging the challenges and concerns that have arisen as a result of its rapid growth and expansion. Finally, we will offer our final thoughts and recommendations for the future of infotainment, charting a path forward that embraces innovation and ensures the continued evolution of this captivating field.

A Comprehensive Recap of the Infotainment Landscape

As we reach the conclusion of our exploration into the world of infotainment, it is essential to take a step back and reflect on the major themes and findings that have emerged throughout this book. By doing so, we can better understand the infotainment landscape and its implications on modern society.

One of the most prominent themes that has surfaced is the convergence of information and entertainment. In today's fast-paced digital age, consumers are increasingly seeking content that is not only informative but also engaging and entertaining. This has led to the rise of various infotainment formats, such as edutainment, docudramas, and news satire shows, which blend factual information with creative storytelling and humor. This convergence has transformed how we consume information and challenged traditional boundaries between journalism, education, and entertainment.

Another key finding is the role of technology in driving the infotainment revolution. Advancements in digital media, streaming platforms, and social networking sites have made it easier than ever for content creators to produce and distribute infotainment to a global audience. These platforms have also enabled greater interactivity and personalization, allowing users to engage with content in new and immersive ways. As a result, the infotainment industry has experienced rapid growth and diversification, with an ever-expanding array of content catering to diverse interests and demographics.

The democratization of content creation and distribution is another significant theme that has emerged. With the proliferation of affordable and user-friendly digital tools, individuals and small-scale creators now have the ability to produce high-quality infotainment content without the backing of major media corporations. This has led to a more diverse and inclusive infotainment landscape, with a wider range of voices and perspectives being represented.

Throughout this book, we have also examined the various ways in which infotainment can be both beneficial and detrimental to society. On the one hand, infotainment has the potential to educate, inform, and inspire, making complex topics more accessible and engaging for a wider audience. On the other hand, the blurring of lines between fact and fiction, as well as the prioritization of entertainment over accuracy, can lead to misinformation and a superficial understanding of important issues.

Finally, the ethical considerations surrounding infotainment have been a recurring theme. As content creators and consumers, we must be mindful of the potential consequences of our actions and strive to strike a balance between entertainment and responsible information dissemination. This includes being transparent about the sources of information, acknowledging biases, and promoting media literacy among audiences.

In summary, the infotainment landscape is a complex and ever-evolving phenomenon, shaped by the convergence of information and entertainment, technological advancements, and the democratization of content creation. As we continue to navigate this dynamic terrain, we must remain aware of the potential benefits and pitfalls of infotainment and our ethical responsibilities as creators and consumers.

The Far-reaching Effects of Infotainment on Modern Society

The infotainment revolution has undeniably transformed the way we consume information and entertainment, leaving a lasting impact on various aspects of modern society. This section delves into the implications and significance of infotainment, exploring its far-reaching effects on our daily lives, culture, and global communication.

One of the most significant implications of infotainment is the democratization of information. With the advent of the internet and social media, infotainment has become more accessible to the masses, breaking down barriers that once limited the flow of information. This has led to a more informed and engaged public, as people from all walks of life can now access news, entertainment, and educational content at their fingertips.

Infotainment has also played a crucial role in shaping public opinion and discourse. As the line between news and entertainment blurs, audiences are increasingly drawn to content that is both informative and engaging. This has led to the rise of opinion-driven journalism, where news outlets and content creators cater to the interests and biases of their target audience. While this approach has undoubtedly increased audience engagement, it has also raised concerns about the potential for echo chambers and the erosion of objective journalism.

Another significant effect of infotainment is its influence on cultural globalization. As infotainment content transcends geographical boundaries, it fosters a shared global culture that transcends national borders. This has led to the proliferation of international collaborations, cross-cultural exchanges, and the spread of ideas, values, and beliefs across the world. However, this cultural convergence has also sparked debates about cultural homogenization and the potential loss of local and indigenous cultures.

The infotainment revolution has also profoundly impacted the economy, particularly in the media and entertainment industries. The demand for engaging and informative content has led to the growth of new media platforms, job opportunities, and revenue streams. This has also spurred innovation in content creation, distribution, and marketing, as companies and individuals strive to stay ahead in the competitive infotainment landscape.

Despite its numerous benefits, the infotainment revolution has also raised concerns about the quality and credibility of information. The rise of sensationalism, clickbait, and misinformation has led to a growing distrust of media sources, making it increasingly difficult for audiences to discern fact from fiction. This has significant implications for public discourse and the functioning of democratic societies that rely on an informed and discerning citizenry.

In conclusion, the far-reaching effects of infotainment on modern society are both profound and multifaceted. While the democratization of information, cultural globalization, and economic growth are undeniably positive outcomes, the challenges of sensationalism, echo chambers, and misinformation cannot be ignored. As we continue to navigate the infotainment revolution, we must strike a balance between engagement and credibility, ensuring that the content we consume is both entertaining and informative while also fostering critical thinking and discernment among audiences.

Addressing the Challenges and Concerns in the Infotainment Industry

As we have explored the vast landscape of infotainment throughout this book, it is essential to acknowledge that the industry is not without its limitations and critiques. In this section, we will address some of the most pressing challenges and concerns that have emerged as infotainment continues to evolve and reshape our modern society.

One of the primary concerns in the infotainment industry is the blurring of lines between news and entertainment. With the increasing demand for engaging and captivating content, there is a risk that the integrity of journalism may be compromised. News organizations may prioritize sensationalism and clickbait over accuracy and objectivity, leading to a decline in the quality of information available to the public. This phenomenon has given rise to the term "infotainment," which critics argue is a dangerous mix of information and entertainment that can mislead and manipulate audiences.

Another challenge faced by the infotainment industry is the potential for echo chambers and filter bubbles. As consumers increasingly rely on personalized content feeds and social media algorithms, they may become trapped in a self-reinforcing cycle of information that confirms their existing beliefs and biases. This can lead to a fragmented and polarized society, where individuals are less likely to engage in constructive dialogue and debate.

Moreover, the infotainment industry has been criticized for its role in perpetuating stereotypes and reinforcing harmful cultural norms. By prioritizing entertainment value, content creators may inadvertently perpetuate harmful stereotypes and marginalize minority voices. This can have a detrimental impact on society, as it perpetuates ignorance and intolerance.

The commercialization of infotainment also raises concerns about the influence of advertisers and corporate interests on content. As media organizations increasingly rely on advertising revenue, there is a risk that editorial decisions may be influenced by the need to appease sponsors and generate profits. This can lead to a decline in the diversity and quality of content available to audiences and a potential erosion of journalistic independence.

In light of these challenges and concerns, it is crucial for the infotainment industry to engage in self-reflection and strive for continuous improvement. Content creators, journalists, and media organizations must strike a delicate balance between providing engaging content and upholding the principles of accuracy, objectivity, and diversity. By acknowledging and addressing these limitations and critiques, the infotainment industry can continue to evolve and innovate while maintaining its commitment to informing and entertaining audiences responsibly and ethically.

Charting the Path Forward for Infotainment Innovation

As we conclude our exploration into the world of infotainment, it is essential to reflect on the insights gained and consider the future trajectory of this dynamic industry. The infotainment revolution has undoubtedly transformed the way we consume information and entertainment, creating a seamless blend of the two that caters to our ever-evolving digital lifestyles. However, as with any innovation, there are challenges and concerns that must be addressed to ensure the continued growth and success of the infotainment industry. In this final section, we will outline some recommendations for charting the path forward for infotainment innovation.

First and foremost, infotainment creators and providers must prioritize the quality and accuracy of the content they produce. In the age of misinformation and fake news, the responsibility of delivering reliable and trustworthy information cannot be overstated. Infotainment providers must invest in rigorous fact-checking and verification processes to maintain their credibility and uphold the integrity of the industry. Furthermore, striking the right balance between entertainment and information is essential to ensure the audience remains engaged while gaining valuable knowledge.

Secondly, embracing diversity and inclusivity in the infotainment industry is vital for its continued growth and relevance. As the world becomes increasingly interconnected, it is essential to represent a wide range of perspectives, cultures, and experiences in the content we consume. Infotainment providers must actively seek out and promote diverse voices, both in front of and behind the camera, to create a more inclusive and representative media landscape. This fosters a greater sense of empathy and understanding among audiences and enriches the overall quality of the content.

Another recommendation for the infotainment industry is to harness the power of emerging technologies to enhance the user experience. Virtual reality, augmented reality, and artificial intelligence offer exciting possibilities for creating immersive and interactive infotainment experiences that captivate audiences and deepen their engagement with the content. By staying at the forefront of technological advancements, infotainment providers can continue to innovate and push the boundaries of what is possible in this rapidly evolving field.

Moreover, the infotainment industry needs to prioritize ethical considerations and social responsibility in their content creation and dissemination. This includes addressing issues such as privacy concerns, data security, and the potential for manipulation or exploitation of vulnerable audiences. Infotainment providers must establish and adhere to robust ethical guidelines that protect the interests of their users and promote a safe and responsible media environment.

Lastly, fostering collaboration and partnerships within the infotainment industry can help drive innovation and ensure its long-term success. By working together, content creators, technology providers, and other stakeholders can pool their resources, expertise, and ideas to develop groundbreaking infotainment experiences that cater to the diverse needs and preferences of audiences worldwide.

In conclusion, the infotainment revolution has irrevocably changed the way we consume information and entertainment, offering exciting new possibilities for engaging and educating audiences. By addressing the challenges and concerns accompanying this innovation and embracing the recommendations outlined above, the infotainment industry can continue to thrive and evolve, enriching our lives and shaping the future of media consumption.

ABOUT THE AUTHOR

Kiannah Frost is a prolific author and thought leader in the fields of psychology, technology, and social dynamics. With a keen eye for understanding the complexities of human behavior, she has dedicated her career to exploring the ways in which people interact with one another, both in person and through various forms of media.

Kiannah's first series, Mindful Momentum, delved into the power of mindfulness and its impact on personal growth and relationships. Her second series, Breaking Stereotypes, challenged societal norms and encouraged readers to embrace their unique identities. Now, with her third series, Navigating the Digital Age, Kiannah turns her attention to the rapidly evolving world of technology and its influence on human connection.

When she's not writing or coaching, Kiannah enjoys spending time in nature, practicing yoga, and traveling the world, always seeking out new experiences and insights to share with her readers.

Made in the USA
Monee, IL
04 November 2026

67379451R00079